Nature, Temporality and Environmental Management

How are different concepts of nature and time embedded into human practices of landscape and environmental management? And how can temporalities that entwine past, present and future help us deal with challenges on the ground? In a time of uncertainty and climate change, how much can we hold onto ideals of nature rooted in a pristine and stable past? The Scandinavian and Australian perspectives in this book throw fresh light on these questions and explore new possibilities and challenges in uncertain and changing landscapes of the future.

This book presents examples from farmers, gardens and Indigenous communities, among others, and shows that many people and communities are already actively engaging with environmental change and uncertainty. The book is structured around four themes – environmental futures, mobile natures, indigenous and colonial legacies, heritage and management. Part I includes important contributions towards contemporary environmental management debates, yet the chapters in this section also show how the legacy of older landscapes forms part of the active production of future ones. Part II examines the challenges of living with mobile natures, as it is acknowledged that environments, natures and people do not stand still. An important dimension of the heritage and contemporary politics of Australia, Sweden and Norway is the presence of Indigenous peoples. As is clear in Part III, the legacies of the colonial past both haunt and energise contemporary land management decisions. Finally, Part IV demonstrates how the history and heritage of landscapes, including human activities in those landscapes, are entwined with contemporary environmental management.

The rich empirical content of the chapters exposes the diversity of meanings, practices and ways of being in nature that can be derived from cultural environmental research in different disciplines. The everyday engagements between people, nature and temporalities provide important creative resources with which to meet future challenges.

Lesley Head is Redmond Barry Distinguished Professor and Head of the School of Geography at the University of Melbourne, Australia.

Katarina Saltzman is Associate Professor at the Department of Conservation, University of Gothenburg, Sweden.

Gunhild Setten is Professor of Geography at the Department of Geography, Norwegian University of Science and Technology, Trondheim, Norway.

Marie Stenseke is Professor of Human Geography at the Department of Economy and Society, University of Gothenburg, Sweden, and Fellow of The Royal Swedish Academy of Agriculture and Forestry.

Nature, Temporality and Environmental Management

Scandinavian and Australian perspectives on peoples and landscapes

Edited by
Lesley Head, Katarina Saltzman,
Gunhild Setten and Marie Stenseke

Taylor & Francis Group

LONDON AND NEW YORK

First published 2017 by Routledge

2 Park Square, Milton Park, Abingdon, Oxfordshire OX14 4RN
52 Vanderbilt Avenue, New York, NY 10017

Routledge is an imprint of the Taylor & Francis Group, an informa business

First issued in paperback 2020

British Library Cataloguing in Publication Data
A catalogue record for this book is available from the British Library

Library of Congress Cataloging in Publication Data
Names: Head, Lesley, editor.
Title: Nature, temporality and environmental management : Scandinavian and Australian perspectives on peoples and landscapes / edited by Lesley Head, Katarina Saltzman, Gunhild Setten and Marie Stenseke.
Description: Abingdon, Oxon ; New York, NY : Routledge, 2017.
Identifiers: LCCN 2016009915| ISBN 9781472464651 (hardback) | ISBN 9781315597591 (e-book)
Subjects: LCSH: Human ecology—Scandinavia. | Human ecology—Australia. | Environmental management—Scandinavia. | Environmental management—Australia. | Landscape assessment—Scandinavia. | Landscape assessment—Australia.
Classification: LCC GF611 .N37 2017 | DDC 304.20948—dc23
LC record available at https://lccn.loc.gov/2016009915

ISBN: 978-1-4724-6465-1 (hbk)
ISBN: 978-0-367-66815-0 (pbk)

Typeset in Times New Roman
by Book Now Ltd, London

Contents

Illustrations

Figures

Tables

Contributors

Editors

Lesley Head is Redmond Barry Distinguished Professor and Head of the School of Geography at the University of Melbourne, Australia. Most of this book was completed while she was Director of the Australian Centre for Cultural Environmental Research at the University of Wollongong. She has long-term research interests in human–environment relations and interactions over different timescales. Lesley has been King Carl XVI Gustaf Visiting Professor of Environmental Science at the University of Kristianstad, Sweden (2005–6) and also Visiting Professor in the Department of Economy and Society at the University of Gothenburg, Sweden (2012–14). In 2015 she was awarded the Vega Medal by the Swedish Society for Anthropology and Geography.

Katarina Saltzman is Associate Professor at the Department of Conservation, University of Gothenburg, Sweden. She has a PhD in ethnology and has for many years been researching landscapes and nature/culture relations from an ethnological viewpoint, in close collaboration with scholars from other disciplines. Her research includes rural and agricultural landscape management as well as urban and semi-urban environments, including transitory and temporarily leftover places in the urban fringe as well as intensively tended private gardens. In all of these contexts, the aspect of heritage making has been one of her specific interests.

Gunhild Setten is Professor of Geography at the Department of Geography, Norwegian University of Science and Technology, Trondheim. She has extensive experience from research on human–nature relations, illustrated by outdoor recreation, agricultural practices, cultural heritage and ecosystem services. She has also undertaken research to advance the conceptual development of 'landscape'. Setten is currently leading a project on 'Natural hazards and climate change: the geography of community resilience in Norway', including working with colleagues across the Nordic countries within the recently established Nordic Centre of Excellence on Resilience and Societal Security.

Marie Stenseke is Professor in Human Geography at the Department of Economy and Society, University of Gothenburg, and Fellow of The Royal Swedish Academy of Agriculture and Forestry. She has done extensive research on landscape management and planning in Sweden as well as internationally concerning biodiversity maintenance, comanagement, outdoor recreation, comprehensive planning and conceptual issues. The main part of her work has been carried out in interdisciplinary programs and projects. Stenseke is currently involved in a number of projects on the integration of nature and culture in landscape management, with a certain attention to the complicating issue of temporality.

Contributors

Michael Adams is a geographer at the Australian Centre for Cultural Environmental Research at the University of Wollongong (UOW), Australia. His publications have examined relationships between Indigenous peoples and conservation agencies in Australia and Sweden, including World Heritage, national parks and tourism. Recent research examines hunting in Australia and India, the significance of degraded habitats, and cultural relationships with animals. He has 20 years of employment and consultancy experience with Aboriginal and environment organisations prior to joining UOW. He is a member of the World Commission on Protected Areas of IUCN.

Mattias Åhrén is a professor at the Faculty of Law at UiT – the Arctic University of Norway, where he received his PhD in 2010. Åhrén holds a Master of Law degree from the Universities of Stockholm and Chicago. He has written extensively on Sami and Indigenous rights and lectures on Indigenous peoples' rights around the world. Åhrén also has extensive experience with practical work within the sphere of Indigenous peoples' rights, particularly within the UN. He was a member of the expert group that negotiated the draft Nordic Sami Convention.

Ruth Beilin (Faculty of Science, University of Melbourne, Australia) has more than 30 years of experience in practice-based, community-centred natural resource management and leads an interdisciplinary research lab at the practice–policy interface.

Benjamin Cooke is a lecturer in Sustainability and Urban Planning at RMIT University in Melbourne, Australia, where he currently teaches courses in Environmental Planning and Environmental Theory. He has professional and research experience in environmental management, rural landscape change, land use policy and private land conservation. Benjamin's research interests include human–environment relations and nonhuman agency, the temporality of novel and emerging ecologies, environmental management in the Anthropocene and ecological commoning practices.

Aidan Davison is a senior lecturer in Human Geography and Environmental Studies at the University of Tasmania. The author of *Technology and the Contested Meanings of Sustainability* (SUNY Press) and over 60 journal articles and book chapters, Aidan is fascinated and troubled by interdisciplinary questions of sustainability that arise at the intersection of themes of nature, culture and technology. His theoretical and qualitative research has covered topics such as suburban history, urban planning, urban forest, environmental movements, human–animal and human–plant relations, lay engagement with climate change and educational approaches to sustainability.

Simon Jakobsson is a PhD candidate in landscape ecology at the Department of Physical Geography, Stockholm University. His research focuses on the effects of management on biodiversity across various spatial and temporal scales. His current PhD project investigates the effects of EU policy recommendations on biodiversity in woody pastures in southern Sweden. He has also been involved in research projects in alpine regions of Sweden as well as in agricultural landscapes of Japan.

Dolly Jørgensen, a historian of the environment and technology, has broad research interests ranging from medieval to modern history. Her research areas include modern environmental policy related to ecological restoration and conservation, medieval agriculture and urban sanitation, offshore oil policy, and environmentalism in science fiction. She was a practising environmental engineer before earning a PhD in history from the University of Virginia, USA, in 2008. She is currently an associate professor at Luleå University of Technology, Sweden.

Rebecca Lawrence is a research fellow at the Department of Political Science at Stockholm University. Her research is interdisciplinary and focuses on the intersection of Indigenous claims with governments and the private sector. Her current research projects concern relations between the mining industry and local/Indigenous communities in Sweden, Finland, Norway, South Africa and Australia. Central to these research concerns are the rights of Indigenous peoples, ethical responsibilities of the corporate sector and the chain of accountability throughout global resource industries. Rebecca works closely with companies, NGOs and communities in her research.

Marianne E. Lien is a professor at the Department of Social Anthropology, University of Oslo and began her anthropological career with fieldwork in Finnmark, Norway, in the mid-1980s. She has also done fieldwork on salmon aquaculture in Tasmania and West Norway. Her current fieldwork interests involve food and landscape practices and user rights in coastal Finnmark. Lien has published extensively on food consumption and production, marketing, nature engagement, invasive species, domestication and salmon aquaculture. She leads

'Arctic domestication in the era of the Anthropocene' at the Centre for Advanced Research. Her most recent book is *Becoming Salmon; Aquaculture and the Domestication of a Fish.*

Regina Lindborg has a PhD in Plant Ecology and holds a position as professor in Landscape Ecology at the Department of Physical Geography, Stockholm University. Her primary research field is biodiversity conservation in agricultural landscapes and effects of management and land use change at large spatial and temporal scales. She is currently involved in several projects addressing proactive management for farmland biodiversity in Sweden, Australia, Portugal and South Africa, with focus on the role of life history traits and functional diversity for better understanding of ecosystem processes and ecosystem service generation through time and space.

Libby Robin, Fellow of the Australian Academy of Humanities, is Professor of Environmental History at the Fenner School of Environment and Society, Australian National University, and Affiliated Professor, KTH Royal Institute of Technology, Stockholm, and National Museum of Australia Canberra. She works closely on how museums represent the Anthropocene. Her recent books include the prize-winning anthology, *The Future of Nature* (co-edited with Sverker Sörlin and Paul Warde; Yale UP 2013) and *Curating the Future* (co-edited with Jennifer Newell and Kirsten Wehner). She has won Australian national prizes in zoology, literature and history.

Mattias Sandberg is Associate Senior Lecturer in Human Geography at the University of Gothenburg with an overall research interest in human–nature relations, landscape planning and political ecology. His thesis concerns children's everyday use and understanding of nearby nature and this work was carried out within an interdisciplinary research project on outdoor recreation.

Carina Sjöholm is Associate Professor in Ethnology at the Department of Service Management and Service Studies, Lund University, Sweden. She has previously researched spatial aspects of popular culture and experience economy. She has more specifically centred on social relations, identity formation, commodification of places and materiality. She is currently focusing on tourist experiences such as literary and film tourism and lifestyle entrepreneurship in the rural area.

Elin Slätmo is a Postdoctoral Fellow at the Department of Urban and Rural Development, Swedish University of Agricultural Sciences. She has a PhD in Human Geography from the University of Gothenburg, Sweden. Her overall research interests are sustainable and long-term use of resources, including developing understandings of how our world is constantly changing and why these changes are happening, but also questions concerning how people value and use the physical environment. In her PhD she theoretically scrutinised and empirically studied the driving forces for agricultural land use change in Sweden and Norway.

The postdoctoral research project focuses on what 'sustainable agriculture' is in the Scandinavian context.

Gro B. Ween is Associate Professor at the Cultural History Museum at the University of Oslo and also the keeper of its Arctic and Australian collections. She started off her anthropological career working in the Australian North, on topics of Indigenous politics, negotiations over rights, landscapes, history and identity. Upon returning to Norway, Ween has applied her fieldwork on Southern Sami, as well as Sea Sami and River Sami in Finnmark. Ween has contributed to the Finnmark Commission, but also written on topics of enactment of rights, negotiations over landscape and nature practices, articulations of culture and of ongoing work to publically establish ontological and epistemological difference in relation to non-Sami Norwegian society.

Simon West is a PhD candidate in Natural Resource Management at the Stockholm Resilience Centre, Stockholm University, Sweden. Simon's research explores knowledge politics, practices and performances in conservation and environmental management, with a current focus on Australia and South Africa.

Stewart Williams is Senior Lecturer in Human Geography and Environmental Planning at the University of Tasmania, Australia. His research on the geographies of risk, regulation and resilience combines an understanding of critical social and political theory with expertise in qualitative modes of inquiry. Stewart has published widely on such pressing issues as drugs and drug use, natural hazards, climate change adaptation and housing and homelessness. He has worked with local communities on matters ranging from public health service provision to sustainable transport infrastructure development.

Introduction

1 Holding on and letting go

Nature, temporality and environmental management

Lesley Head, Katarina Saltzman,
Gunhild Setten and Marie Stenseke

A politics of letting go (or not)

This book explores how different concepts of nature and time are embedded in diverse human practices of landscape and environmental management. Our ambition is to discuss a variety of temporalities, including perspectives that look backward to questions of history and heritage, and forward to the Anthropocene and the uncertainties of climate change. The book examines how these multiple temporalities are entwined with human environmental interactions, landscape management and nature conservation in the present.

The chapters in this book add some embodied examples to the questions of how people engage with nature, and how such engagements transform both their environmental understandings and the material landscape itself. Such capacities provide important cultural resources in rapidly changing circumstances. Understanding and reframing the ways in which humans are conceptualised in relation to the rest of nature is an essential part of working out how to craft a more sustainable future. Incorporating research traditions from both the social sciences and humanities with those of the natural sciences is vital to better understand the distributive, political and cultural dimensions of global environmental problems (cf Hornborg 2009; Head and Stenseke 2014).

Research traditions attending to cultural dimensions expose how people relate to confusing and uncertain (abstract) futures while hanging on to various pasts through the reproduction of landscapes by means of embodied and other (concrete) practices. At the same time, embodied practices also have the potential to generate transformed knowledge and governance processes. Stable territorial and temporal solutions relating to the environment, such as the establishment of protected areas, are increasingly challenged, including by the movement of species. It is worth considering the extent to which the reproduction of landscapes is more common in 'rural' or less densely settled areas, somehow related to a nostalgic view of past environments, whereas we are more used to the 'urban' as a site of ongoing change (Jones 2011).

We live in a time of rapid changes to the global system, and it is anticipated that future changes will accelerate. It is also increasingly accepted that humans are now the dominant force affecting Earth surface processes – hence the encapsulation of the current age as the Anthropocene. Even though we want to warn against seeing the Anthropocene as a 'brave new world', the Anthropocene appears to be a place and time of spatial and temporal 'cross-fire' where past and future, local and global are mobilised and come together to create new entanglements which are characterised by uncertainty, loss of control and risk. Yet people mobilise different environments, through their (embodied) practices and institutional strategies, to keep and regain control, to fix whatever is lost and to produce predictable futures, as is demonstrated in several contributions in this volume. For that, we need a past which can aid in 'efforts to know what lies ahead' (Adam and Groves 2007, p. 6). At any time, then, continuity proves paramount and values developed over hundreds of years are heavily influencing both our abilities and will to 'face the future, rather than running from the past' (Lorimer 2015, p. 4).

In her contribution in this volume, Dolly Jørgensen states that 'humans of the twentieth and twenty-first centuries have shown a remarkable unwillingness to let go' (p. 55). Letting go – or, alternatively, hanging on to a past – involves grief and mourning, increasingly evident in conservation thinking (Robbins and Moore 2013; Head 2016). Such grief echoes the punctuation of a grand narrative of Nature: there appears to be no monolithic story about Nature anymore. Indeed recent social scientific scholarship tells us there has never been one (Castree 2015), yet the modern figure of Nature – to which we repeatedly turn, and which is 'known by objective Science and defended and restored by rational environmental management' (Lorimer 2015, p. 2) – continues to set the agenda for global environmental policies as well as for practices in home gardens. This tension is evident in various international environmental frameworks; for example, the Intergovernmental Platform on Biodiversity and Ecosystem Services (IPBES) continues to distinguish between natural and human processes, but also attempts to embrace different worldviews (Díaz *et al.* 2015).

Concurrently, on the other hand, there are new possibilities in a world characterised by non-linearity, dissonance and multiplicity. In some situations there is relief that the master narrative of a single Nature has fragmented. New environmental practices can come to the fore, in addition to a range of vernacular and culturally diverse practices not fully recognised under environmental management regimes so strongly framed by scientific thinking. In this book we present everyday examples from Indigenous communities, farmers and gardeners, among others. Our argument is not that the past must be abandoned but that a more granular analysis of how different temporalities are entwined is necessary.

Why Scandinavian and Australian perspectives?

Increasingly unstable environments generate and may require increasing mobility – of people, other species and ideas. This book has emerged from a particular traffic in academic mobility between Australia, Sweden and Norway over the last decade or so. The book celebrates and draws on the considerable scholarly interchange between the three areas, as represented in the list of contributors. This has led to collaborative projects and workshops, reciprocal visits, postgraduate student exchanges and joint publications.

The notion of mobility is also relevant to our themes. Environments, natures and people do not stand still. Their mobility is both spatial and temporal, as seen, for example, in debates over alien species. Landscapes are therefore not only about roots and locatedness, they are just as much about movement and permeable scalar relationships. In understanding relationships between nature, temporality and environmental management, we simply cannot afford to risk 'losing sight of the larger picture, the larger fields of political, social and economic power and processes of change' (Bender 2001, p. 83).

We have chosen to combine and juxtapose perspectives from the different regions to see where this might lead. Our contributor Libby Robin argues that these perspectives are 'off-centre', albeit highly influential. They thus shed a different light on pervasive issues debated through the affluent West – the human role in landscape management, the significance of protected areas for biodiversity conservation, culturally diverse perspectives on nature, and the importance of tangible and intangible heritage in landscapes subject to social and ecological change. In each of our focus countries, local voices have challenged the dominant national environmental imaginary, and vernacular ways of living with nature (including those of both Indigenous peoples and farmers) have made important contributions to understanding and practice. New collaborative land management arrangements are emerging.

The comparisons between Sweden, Norway and Australia are not obvious, but we have found them instructive. Despite considerable differences when it comes to physical geography, climate and history, there are in fact a number of relevant similarities between the countries. They all have advanced economies and are relatively sparsely populated. In Sweden, population is concentrated in the south; in Australia it is concentrated around the coastal fringe, particularly of the southeast; whereas in Norway, even though the majority of the population lives in the south, a much more scattered settlement pattern is found. This leaves in each country significant areas of remote country (variously arctic and arid/tropical) for more extensive land uses including pastoralism, forest management, national parks and Indigenous land. In each country, farming has an important place in the national biography and in national identity, running in a somewhat parallel narrative to the valuation of 'wild' nature. As in many Western countries, there is anxiety about the processes of rural decline in marginal areas, with decreasing rural populations and weakening social networks in rural communities. In all countries there are narratives about growing distances (both conceptual and social) between city and country. At the same time there is an increasing difference between the peri-urban countryside and more remote areas, in terms of land use interests, with a counter-trend of amenity migration to rural areas accessible to large cities.

There are also important differences in environmental management and biodiversity protection. For example, in Sweden, at least for grassland ecosystems, environmental protection can occur as part of farming; in Australia it must usually be done in spite of it (Saltzman *et al.* 2011). We have categorised these patterns as integrationist and separationist, respectively. These patterns should themselves be considered historically contingent, and they give us some inklings of future possible trajectories. Four of the chapters take an explicitly comparative perspective (Robin, Adams, Slätmo, Jørgensen), but there are many links that can be drawn.

The contributors to the book come from diverse disciplinary backgrounds, including geography, environmental history, anthropology, political science and ethnology. They have used a diversity of research methods (ethnography, survey, in-depth interview, observations of various kinds, documentary and archival analysis) in a variety of contexts (suburban gardens, farms, hunting, peri-urban landscapes and remote mining landscapes). The point is not to derive generalisations, but to expose the richness and diversity of meanings, practices and ways of being in nature that can be derived from cultural environmental research (Head and Stenseke 2014). Further, nature and landscape are mobilised in different contexts for different ends, exerting many kinds of agency.

It is striking how deeply the colonial pasts of all three nations continue to both haunt and energise the present (e.g., chapters by Ween and Lien, Lawrence and Åhrén). Even in chapters that are not explicitly about Indigenous land management issues, as Davison and Williams's chapter shows, the colonial heritage pervades contemporary decision-making and is embedded in many aspects of the landscape. Such issues have been much more a part of the national conversation in Australia than in Sweden and Norway. These are difficult issues to discuss, more so in Sweden it seems than Australia (see, e.g., Lawrence and Åhrén's chapter). In the preparation of this book, we found ourselves wondering whether we have different understandings of even what colonialism means.

Part I: imagining new environmental futures and entwined pasts

There are many Anthropocenes in contemporary discourse, reflecting both conceptual challenges and the unpredictability of various climate change scenarios. Nevertheless the diagnosis of humans as geological actors at a global scale is revolutionary. 'Unsurprisingly perhaps, for some publics the magnitude and consequences of our geological entanglements are proving hard to accept' (Lorimer 2015, p. 1), not least because they challenge Nature as a single and atemporal domain apart from society.

In Part I, we discuss environmental futures, with important contributions towards contemporary environmental management debates. Yet the chapters in this section also show how deep understandings of past landscapes and peoples are embedded in discussions of present and future. In several of these chapters, the legacy of older landscapes forms part of the active production of future ones.

Aidan Davison and Stewart Williams demonstrate clearly how the temporalities of past, present and future are always intertwined. The colonial past continues to be with us, as the space/time of the frontier continues to be played out and invoked in contemporary debate. Providing an exemplary overview of the ontological practices – ways of being in the world – of Australian modernity, they set the scene and provide context for the rest of the book. While it may be challenging to open the book with ontological questions, Davison and Williams make a powerful argument that it is precisely such questions that are fundamental to the Anthropocene. They argue that while the temporalities of Indigenous and modernist ontologies are incompatible, the challenges posed to modernity by the continuing Indigenous present provide some inkling of what the broader set of future challenges might require. They also provide an overview of non-equilibrium ecologies of relevance to all the case studies in the book – the challenge of dealing with both mobility and stasis and, by implication, uncertainty. Contributing to the broader critique of ontology of human exceptionalism, including as reproduced in Anthropocene discourses of human power over the Earth, Davison and Williams offer instead the concept of the 'intimate otherness' of the environment as being integral to the exploration of human possibility.

Gunhild Setten weaves temporality, responsibility and landscape into a conceptual conversation, using examples from invasive alien species debates. (Such species reappear a number of times in the book, particularly in Part II.) She argues that because the past is so embedded in understandings of landscape, combined with the physical landscape's ability to hold on to a past, the concept of the future becomes taken for granted to some extent, and geographers rarely engage with 'the future' as a category. Uncertain future ecologies put explicit pressure on configurations of responsibility, whether legal or moral. Setten raises three challenges or questions in order to problematise the complex relationship between nature, time and responsibility: First, who bears responsibility in different sites, spaces and times of sustainability configurations? Second, what might responsibility mean when we recognise that shifting contexts can introduce unplanned forms of culpability? Third, how does 'landscape' inform a debate about responsibility – on the ground as well as analytically? Together, these might teach us something about the importance of explicitly engaging with a future in order to understand how temporality and responsibility work to (re)produce certain landscapes.

Dolly Jørgensen discusses the temporal horizons of extinction, through examples from two places on the edge or the fringes of civilization – the beaver in Sweden at the end of the nineteenth century, and the thylacine of Tasmania in the twentieth. Both northern Sweden and Tasmania are widely understood as wild and unknown places, where it is entirely imaginable that mysteries about nature can persist. She explores the complex processes by which 'the presence of an absence (no known animals) became understood as an absence of presence (extinction)'. Among other things, the chapter forces us to ponder just why we are so obsessed with the life and death of the last individual of a species, much more than with other individual lives. It also forces us to ponder the following question: 'Can one be sure that something is not there simply because one doesn't see it?' In both the Swedish and Tasmanian cases, argues Jørgensen, the extinction story becomes woven as something that was somehow avoidable, even when down to the last animal. Both stories have strongly political conservation messages, as the extinctions take place very close to the times of species protection. The unwillingness to let go, that we have discussed above, can have direct effects on present nature conservation practices. Since recovery of species comes with a cost while time and money is limited, it also has consequences for what actions are prioritised.

Libby Robin takes a unique approach to the Swedish-Australian comparison, examining the role of the two countries in what she calls 'the imaginative discourse of the Anthropocene'. This discourse embraces diverse activities, from participation in scientific enterprises such as the Stockholm-based International Geosphere Biosphere Program and the Future Earth program, to artistic and cultural activities. Robin uses the lenses of nationhood and science to compare 'two smaller nations anxious to be leaders in a world bent on modernity' (p. 65). The national identities of both Sweden and Australia are at least partly based 'on heroic relations with extreme places' (p. 63). Both have had strong public science programs seeking to make sense of their respective Far Norths and other remote places as part of the nation-building project. Robin charts a brief intellectual history of the environmental humanities, sometimes referred to as the *Ecological Humanities* in Australia, and the *Green Humanities* in Sweden. She argues that they, along with other small, rich nations, have made distinctive contributions to the emergent global debate around the Anthropocene.

Part II: living with nature in motion

The mobilities mentioned disrupt some longstanding patterns, but also allow new possibilities to emerge – more fluid and hybrid modes of living. The chapters in this section show the importance of embodied knowledge in disrupting fixed categorisations and producing new environmental knowledge. This section aims to promote positive solutions as well as identify old conflicts. In these chapters, embodied interactions with non-human others challenge and rework categories of belonging.

Benjamin Cooke examines how rural-amenity landholders both engage with the legacy of past interventions in the landscape and create new ones. The legacies of past plants, animals and soil help structure future trajectories. These landholders often start with what Cooke calls 'redemptive aspirations', such as wanting to plant or replant native species in degraded landscapes. Through a series of examples, Cooke charts the complexities of actually doing this on the ground, such that a number of his participants adopted a philosophy of 'do a little bit and see what happens'. In a theme that is also pursued in the two chapters that follow, Cooke writes about the embodied practice and knowledge that are involved as people interact with plants. This fine-grained study speaks to bigger questions concerning environmental management. Cooke suggests in his chapter that attending to non-human agency can open up opportunities to think differently about ecological change.

Michael Adams's chapter on hunting uses a comparative lens between Australia and Sweden to good effect. Different legal frameworks use different species categories in the two places. Hunting is enrolled differently into national imaginaries. The study provides insights not only into which animals belong but which people and activities belong. The concept of nativeness here takes on a different framing to other chapters where nativeness is discussed. While there is a strong history of rural and outdoors practicality and competence related to hunting in Sweden, and wide acceptance of eating wild foods, Australians might be keen to plant native plants (Cooke) but are estranged from native foods. Adams uses his own embodied knowledge – first-hand encounter in the process of being a hunter – to take the reader into very intimate relations with the co-presences in the landscape. Part of Adams' argument is that bush competence – embodied knowledge, built over time through physically strenuous practice – is considered by hunters 'as evidence of their belonging'. Further, this belonging is fundamentally about the fact that they hunt for food, a kind of visceral 'embodiment of the wild'.

Embodied interactions with non-human others are also to the fore in the chapter by Katarina Saltzman and Carina Sjöholm, who examine the home garden as a site for environmental interactions and for contemporary understandings and constructions of nature and temporality. As the home garden is an environment to which more than half the Swedish population have direct access (a comparable figure for Australia is more than 80 per cent), it is an important site to study if we want to understand everyday environmental management. Saltzman and Sjöholm remind us that 'the concept pair nature/culture is a strong figure of thought' for many people, notwithstanding its deconstruction in academic analysis, but they also argue that 'the complex, dynamic, multi-species conglomerate of the garden can easily be interpreted in terms of hybridity, fluidity and boundary-crossing biosocial becomings' (p. 126). Here they are referring to anthropologist Tim Ingold (2013), who has discussed ensembles of relations and trajectories of lives, which are all simultaneously social and biological.

Part III: Indigenous challenges to environmental imaginaries

An important dimension of the heritage and contemporary politics of Australia, Sweden and Norway is the presence of Indigenous peoples. As is clear in the next two chapters, this has led to much contestation with the nation-states. Without downplaying the conflicts involved, here we try to go further by exploring the possibilities that arise when multiple natures are considered.

Gro Ween and Marianne Lien explore the postcolonial process of establishing local and Indigenous user rights in northern Norway and, in this context, discuss the possibility for articulation of multiple natures. This chapter analyses legal and bureaucratic procedures and uses parallel processes in Australia as a comparative figure, while looking for possible *uncommons* – alliances that incorporate constitutive differences (de la Cadena 2015). 'To what extent is there currently a space for articulating divergences that undo a singular nature?' ask Ween and Lien. While postcolonial ambitions are far-reaching, the authors show that there are a number of hindrances built into the new procedures for participation. In spite of the efforts to include Sami and other local perspectives, the 'lack of open articulation of ontologically divergent framings of what nature, people, and belonging is about' constitutes an obstacle. According to Ween and Lien, this process can even be interpreted as a 're-colonisation in the name of de-colonisation' (p. 144).

Rebecca Lawrence and Mattias Åhrén use the important example of mining to deal with Sweden's contested colonial heritage and explore how colonialism continues to pervade the present. They contrast this with Australia where, as Davison and Williams also argued, questions of colonial history and Indigenous rights are more explicit in national life, albeit emerging more strongly at different times. Lawrence and Åhrén argue that Indigenous rights in Sweden have not been extinguished, but have been made invisible instead. To render them visible in an innovative way, they provide a counterfactual calculation of the amount of mining royalties that would have been paid to Sami in a year had their rights been recognised.

Part IV: temporalities of environmental management

The chapters in this final part demonstrate the history and heritage of landscapes and human activities in those landscapes. They show how decision making about landscapes is proceeding in diverse contexts and how fixed categories of land use are challenged by the more complex realities on the ground.

Elin Slätmo takes as her point of departure that throughout the postwar period there has been a steady decrease in farmland in Sweden and Norway, as well as in other European countries. Reasons for this decrease have commonly been analysed through an operationalisation of differing driving forces set on describing changes in the physical landscape. Slätmo argues that in order to understand land use changes more fully, farmers' agency need to be acknowledged and accounted for. By building on 'key driving forces', that is, physical environment, socioeconomic environment, technology, culture and policy, Slätmo develops drivers that limit, enable and directly trigger farmers' decision making, which to a larger degree can analytically mobilise their agency. According to Slätmo, and through her studies in Norway and Sweden, contextual and, by default, temporal and spatial concerns will help explain whether land is taken out of agricultural production or not.

Ruth Beilin and Simon West undertake a critique of adaptive management, in the context of broader critiques of modernist environmental management. They argue that adaptive management's original ambition – 'to engender open-ended processes of social experimentation in complex situations' (p. 191) – remains valuable. However, two things are necessary to fully realise the benefits of adaptive management – recognising both the multiple subjectivities underlying ideas of nature, and the many natures produced through management practice. Beilin and West turn a performative lens onto management, seeing it as a practice 'grounded in doing'. They use two different examples, prescribed burning as a wildfire management tool, and an Indigenous mapping project in Warlpiri Country in Australia's Northern Territory.

Marie Stenseke, Regina Lindborg, Mattias Sandberg and Simon Jakobsson examine semi-natural grasslands, a biotic community whose very name contains the nature/culture dualism, as if there is a continuum between nature at one end and culture at the other. Their chapter focuses on the temporalities of this agricultural landscape, which is rich in both biological and cultural heritage. A key argument of the chapter is that temporal aspects are rarely explicit in policies and management strategies concerning semi-natural grasslands. As Stenseke and her colleagues show, this gives rise to significant challenges in decision making, especially in addressing the future. They ask, for example, what should be secured, maintained or restored (see also Gunhild Setten's chapter). They consider the temporal rhythms of farming, from the daily scale to the issue of succession planning, and examine the role of old trees in these landscapes. One conclusion is that a long-term perspective in nature conservation demands an inter-sectoral approach. That will open a wider range of possible and promising sustainable pathways.

Conclusion

A practical ambition of this book is to acknowledge and celebrate the considerable scholarly interchange between the two areas over the last decade or so, as represented in the list of contributors. The traffic between Sweden, Norway and Australia in disciplines such as geography, environmental history, Indigenous studies, landscape science and heritage has been considerable. Our intellectual ambition is that this particular confluence of ideas and perspectives will contribute in some way to the contemporary challenges of landscape management and to ongoing global environmental initiatives. The challenges before us as earth citizens demand that we draw creatively on diverse ideas and practices. As the contributors to this book show, many such resources are to be found in everyday engagements between people, nature and temporalities.

Acknowledgements

Funding for collaborations between the editors has been provided at various times by the King Carl XVI Gustaf Visiting Professorship in Environmental Science to Lesley Head, Högskolan Kristianstad, the University of Wollongong, University of Gothenburg and the Norwegian University of Science and Technology, Trondheim. Earlier versions of several chapters were presented at the RGS-IBG Conference, London in 2013, in a session called 'Nature, Time and Environmental Management'. We are indebted to Eliza de Vet for her project management and attention to detail, and to Diane Walton for editing assistance.

References

Adam, B. and Groves, C. 2007. *Future Matters: Action, Knowledge, Ethics*. Leiden, The Netherlands: Brill.
Bender, B. 2001. Landscapes on-the-move. *Journal of Social Archaeology*, 1(1), 75–89.
Castree, N. 2015. *Making Sense of Nature: Representation, Politics and Democracy*. Abingdon, England: Routledge.
de la Cadena, M. 2015. Uncommoning Nature. *Apocalypsis*, 22 August. Accessed 16 April 2016. Available at http://supercommunity.e-flux.com/texts/uncommoning-nature/.
Díaz, S., Demissew, S., Carabias, J., Joly, C., Lonsdale, M., Ash, N. *et al.* 2015. The IPBES Conceptual Framework – connecting nature and people. *Current Opinion in Environmental Sustainability*, 14, 1–16.

Head, L. 2016. *Hope and Grief in the Anthropocene: Re-conceptualising Human–Nature Relations.* Abingdon, England: Routledge.

Head, L. and Stenseke, M. 2014. Humanvetenskapen står för djup och förståelse (The humanities and social sciences stand for depth and understanding), in *Hela vetenskapen! 15 forskare om integrerad forskning (The Whole of Science! 15 Researchers on Integrated Research)*, edited by E. Mineur and B. Myrman. Stockholm: Vetenskapsrådet (Swedish Research Council), 26–33. Accessed 16 April 2016. Available in English as 'Seven contributions of cultural research to the challenges of sustainability and climate change', at: http://www.uowblogs.com/ausccer/2014/11/13/seven-contributions-of-cultural-research-to-the-challenges-of-sustainability-and-climate-change/.

Hornborg, A. 2009. Zero-sum world: Challenges in conceptualizing environmental load displacement and ecologically unequal exchange in a world system. *International Journal of Comparative Sociology,* 50, 237–262.

Ingold, T. 2013. Prospect, in *Biosocial Becomings: Integrating Social and Biological Anthropology*, edited by T. Ingold and G. Pálsson. Cambridge, England: Cambridge University Press, 1–21.

Jones, M. 2011. European landscape and participation – rhetoric or reality?, in *The European Landscape Convention: Challenges of Participation*, edited by M. Jones and M. Stenseke. Dordrecht, The Netherlands: Springers, 27–44.

Lorimer, J. 2015. *Wildlife in the Anthropocene: Conservation after Nature.* Minneapolis, MN: University of Minnesota Press.

Robbins, P. and Moore, S.A. 2013. Ecological anxiety disorder: Diagnosing the politics of the Anthropocene. *Cultural Geographies,* 20(1), 3–19.

Saltzman, K., Head, L. and Stenseke, M. 2011. Do Cows Belong in Nature? The cultural basis of agriculture in Sweden and Australia. *Journal of Rural Studies,* 27, 54–62.

Part I

Imagining new environmental futures and entwined pasts

2 The outside within

The shifting ontological practice of the environment in Australia

Aidan Davison and Stewart Williams

Encountering 'the environment'

> It is on the terrain of ontology that many of the urgent ecological battles need to be fought.
>
> (Morton 2013, p. 22)

A paradox lies in the concept of the environment in modern societies. Defined as the surroundings of life, this concept also asserts humanity's essential distance from the world around it. Used to describe the many ways in which humans are wrapped up inside nonhuman realities, this concept nonetheless carries a contrary and largely implicit cultural inheritance of assumed human difference. The epistemologies of detached, universal reason that have shaped much in modern lives depend upon this assumed gap between human subjects and the objectified environments that they endeavour to survey, appropriate and control (Latour 1993). These epistemologies have come in for sustained criticism over the past century (Bernstein 1983). Less well understood are the ways in which such knowledge-making emerges out of ways of being in the world – out of ontological practices – in which modern humanity stands apart from earthly multitudes as the only form of life that cannot be surrounded by others. Moderns are, in essence, surrounded only by themselves (Heidegger 1977). Despite the abstractness of much discussion of ontology, this is truly a stance, a comportment, a practice of being that is every bit as material as it is cognitive.

A powerful new way of talking about the environment has swept through modern societies in the early years of this century. Distinguished by its temporal framing, *contra* the spatial terms in which environments are most often described, the Anthropocene is an epochal narrative that has caught many imaginations. It was first coined by physical scientists to link material description of human Earth-shaping with normative prescription for human planetary stewardship (Crutzen and Stoermer 2000; Steffen *et al.* 2011). Retelling Earth as a human environment, this narrative has spread far and wide, gathering diverse meanings and applications, its passage cleared by concern about anthropogenic climate change (Castree 2014; Johnson *et al.* 2014; Malm and Hornborg 2014). Yet this narrative also owes much to the modern ontology of human exceptionalism that asserts human-ity's irreducible distance from the reality that surrounds it. Seemingly based on evidence of humanity's seamless existence within more-than-human wholes, modernist accounts of the Anthropocene paradoxically reassert humanity's her-metic autonomy by positioning Earth within the human realm. 'The Human Earth' thus appears as an artefact, a human appendage abstracted from a 'natural' past and given over to a future of human choice. This new but familiar modern story of human destiny has the effect of masking the transcendent autonomy, diversity and scale of nonhuman reality (Clark 2011). This is to live inside mirrored horizons, 'within which human agents yearn for others but encounter only themselves' (Davison 2015, p. 2).

We engage the subject of the environment in Australia in the context of critique of the modern ontology of human exceptionalism in this chapter. We do not confine our interest in environmental management to the discrete profession that bears this name, and we investigate the environment as a temporal as well as a spatial sub-ject; as a space/time made through ontological work. We critically engage modern encounters with environments as external objects amenable to, and requiring par-ticular kinds of, objective and predictive management. We challenge Anthropocene narratives that explain environmental crisis as an opportunity for humanity to take charge of the Earth. Through the Australian case, we present the intimate otherness of the environment, especially its rendering through time and space as variously proximate or distant, as integral to the making of human possibility.

In *Hyperobjects*, Timothy Morton (2013) provides insight into the ways in which critique of human exceptionalism necessarily demands critique of modern accounts of ontology. As noted above, part of such critique is to grasp ontology as a material as well as conceptual encounter: as a human stance taken in rela-tion to the real that is bodily, enacted and felt, as well as thought. Human worlds, including the cleaved universe of modernity, are *made* not *found* in a relational coproduction of human and nonhuman possibilities; in 'human-other-than-human entanglements' in which bodies, things, forces and affects arise and are intelligible. For those of us reared within modern ontologies, the proposition that reality comes to be known through acts of mutual encounter rather than our active description of passive objects is difficult. It goes against the grain of modern worlds of practice. More difficult still is the challenge of articulating these encounters in ways that do not assert any distance between humanity and the reality that exceeds it. As Morton states, *'we are always inside'* the objects we seek to describe (2013, p. 17, emphasis in original). Put differently, and aided by Martin Heidegger's (1977) rendition of ontological practice (see Davison 2001), the encounters in and through which humanity becomes human, and thus inhabits a historical world, have no objective inside or outside. This is the meeting of differences that exist only in acts of meeting. This is the creation of seamless but heterogeneous human-other-than-human wholes (Davison 2015). Morton's description of ontology as terrain is thus literal. Ontology is the human practice of encounter *and* the more-than-human terrain in which such encounters are possible. It is not just that the urgent envi-ronmental issues that beset modern societies and prompt talk of an Anthropocene demand ontological inquiry. These issues are themselves an ontological terrain, an environment of human encounter in which human and other-than-human being is made, and can be made differently. In this sense, the environment is an outside within the human condition.

We address the spatial and temporal specificity of Australian modernity, informed by this call to engage with environments as ontological terrain and as sites for the making of human-other-than-human possibilities. Starting with the environment as a practice of colonial displacement, we explore the coupling of nation and nature in Australian history. We chart the transformation of nature from an unintelligible and thus dangerous other to its present status as an endangered and thus desirable other through the space/time of the frontier. We then consider the contemporary implications of the ongoing collapse of modern experience of nature's stability and remoteness, in the ruins of which stories of the Anthropocene are taking hold. Modern technological practices are making modern ontological divisions between society and nature increasingly untenable. In the context of this paradox, we consider ontological possibilities of belonging in Australian environments.

The frontier environment

The colonial project in Australia was one of territorial invasion followed by the violent dispossession of indigenous peoples and the appropriation of their land (Carter 1987) – a violence both fast and slow, proximate and distended, and active still (Nixon 2011). The advance guard of explorers, surveyors, soldiers, convicts, missionaries, administrators and scientists met a landscape empty of temporal and spatial reference points: a reality without markers. This was to confront utterly alien environments that inspired fear, confusion and paralysis. The task of colonisation, then, was to hold this 'upside down' world at arms' length, to render it a stable, discrete object capable of being managed by universal reason. Australian environments demanded occupation, cultivation and improvement in line with modernist sensibilities around how the land and everything on it was to be valued, appropriated and improved, particularly in relation to liberal systems of private property ownership and exchange. The physical setting of Australia was encountered as an inventory of resources lacking any context that would inhibit their being identified, measured, quantified, apportioned and allocated. This was a reality of disconnected components ideally suited to instrumental reordering.

Like many colonial societies, 'Australians' geographical imaginations have been profoundly affected by frontier metaphors' (Howitt 2001, p. 233; see also Rose and Clarke 1997). Embodying this imaginary in practice, frontiersmen and women displaced the radical otherness of antipodean natures and cultures to the other side of an unbridgeable divide, allowing them to settle into a hospitable reality. Holding coherence and incoherence apart, the frontier was inscribed spatially through the material project of civilising wild environments. The movement of this frontier through time was an objective record of progress, a forward motion towards enlightenment as nature made way for modern organisation.

During the nineteenth century, the environment of Australian modernity was a space/time of linear advance, a purposeful arrow of progress written in soil, embodied in plant and animal bodies, expressed with dynamite and axe. A latent reality was possessed and made a coherent and productive world. The inevitable advance of modernity was counterposed to the inevitable decline of Indigenous peoples, who it was assumed would die out as the 'superior' white race took their place. The co-production of Indigenous ways of living and Australian environments was misrecognised. Indigenous cultures were assumed to be constrained by untamed forces of nature. Evidence of their lives was rendered as scientific curiosities already consigned to the past and left to be itemised, documented and deployed in the archives and arguments of anthropologists, phrenologists and eugenicists (Head 2000; Anderson 2008).

As Debbie Bird Rose observes, the 'disjunctive moment' of colonisation cut 'an ontological swathe between "timeless" land and historicised land' (1997, p. 28). The frontier operated as an ontological device for enacting the dualistic metaphysics of modernity. At once cognitive and material, this device generated distance in the context of proximity (Morton 2013). Holding unfamiliar surroundings at bay, the frontier was integral to the production of a space/time in which colonial modernity claimed the right to belong. The threat and the lure of frontiers beyond civilisation have defined much in Australia's modern history. The march of progress through Australian environments and the retreat of the wild frontier to the margins of Australian society is linked to a potent cultural dialectic of triumph and loss. The more the modern enterprise has been inscribed on an apparently blank slate, the more modern Australians have hankered for a past that seems to be slipping away.

The national environment

Understood on its own terms, the progressive establishment of an Anglo-European settler society across Australian environments during the nineteenth century filled the goading vacuum of *terra nullius*, a space perceived to be barren of meaning and without time, pushing the frontier of nature to the margins of colonial society (Lines 1991). The linear history of this society has been mapped through key events and projects explained in terms of geographical expansion (Carter 1987). Modern order was rapidly manifest in antipodean landscapes, from grids of suburban housing to neat lines of wheat, rail and fence. The linear, temporal rhythm of progress was manifest in steady transformation of spaces perceived to be otherwise without purpose. The insensible happenstance of precolonial nature and culture was displaced by linear, and thereby predictive, reason. Modernity moved with remarkable speed and spread (Gascoigne 2002), meeting with less friction from premodern European traditions in Australian environments than was the case in the European homelands in which it arose.

Yet around the turn of the twentieth century, as nationhood was declared and the work of modern reason was widely apparent in Australian landscapes, many in Australian society were yearning for a mythic identity, one with roots beyond modernity (Davison 2005). The world's most suburban nation at this point, modern Australia clung to the coast and reached out towards Britain (Davison 1995). But the gaze of this fledgling nation was arguably on the now distant frontier of nature, where the otherness of precolonial Australian ecologies (and later, precolonial cultures) began to accrue romance and innocence. The cultural archetypes of this suburban nation were those of the drover, prospector and bushman (pastoralist, miner and pioneer), figures who inhabited the distant environments of the outback, the red centre and the bush (Devlin-Glass 1994). Natural history societies flourished, National Parks drew enthusiasts, and distinctively Australian flora and fauna were appropriated as emblems of place-based identity (Franklin 2006). With suburban modernity firmly cemented in Australian soil, the practice of the frontier began to locate a truly distinctive mode of Australian belonging in the space/time of 'the native', a prelapsarian world physically remote and isolated and locked in the moment before colonial invasion (Franklin 2006; Trigger *et al.* 2008; Head 2012).

The national anthem's first verse, for example, asserts the story of industrious pioneers but transmutes earlier experience of nature as a dangerous lack of purpose into an actively generous benefactor endowing Australian life with material abundance and aesthetic meaning:

> We've golden soil and wealth for toil;
> Our home is girt by sea;
> Our land abounds in nature's gifts
> Of beauty rich and rare;
> In history's page, let every stage
> Advance Australia Fair.

This song was composed in 1878 and quickly gained popular recognition and use, but it was only officially instated as the national anthem in 1974 and even then was replaced temporarily with a return to the traditional use of Britain's 'God Save the Queen' between 1976 and 1981 (Australian Government n.d.). Such tensions between the assertion of a benign and welcoming nature in Australia and dreams of recreating an 'old world' in a new place have long confounded a sense of identity and security in the unfolding of the story of Australian nationhood.

Appropriating the biological, geological and hydrological resources of a continent, Australian modernity grew affluent through the twentieth century, with much of this wealth invested in the making of suburban environments. Over this period, the suburban population in the six capital cities alone grew from less than one million to over 11 million (ABS 2003). Interest in natural history and native environments morphed in the second half of this century into ecological sciences and environmental social movements preoccupied with the idea of wilderness as the home of native nature (Lines 2006). In Australia, the idea of wilderness as remote pristine environments with little or no evidence of human habitation owes a debt to colonial narratives of *terra nullius* (Head 2000). To step into this wilderness is to step out of human history. At once sacred and scientific, the figure of wilderness is also a powerful expression of a cultural ambivalence towards the project of modern progress that runs deep in Australian modernity (Davison 2005). In their defence of wilderness, environmental movements manage this ambivalence by shuttling between private affairs of the heart and public arguments of dispassionate reason. More broadly, animated by the dialectic of romanticism and instrumentalism that propels so much in modernity (Latour 2004), wilderness environments are far from remote from the space/time of Australia's suburban heartlands. These suburban worlds promise a liminal zone of contact with modernity in which escape across the frontier into wild nature is always possible (Davison 2005). Predictably, then, the aesthetic-scientific figure of wilderness moved rapidly in the closing decades of the twentieth century into mainstream cultural life and from there into economic logics of recreation, adventure, tourism and place-making.

The ideal of wilderness was entrenched in environmental management practice and on many maps by the 1980s. Around this time, however, the Australian frontier between wild nature and society came under challenge from two different directions, with the result that the figure of wilderness became elusive just as it gained cultural force. First, political recognition of the injustices of colonial invasion called for a new encounter with Australian environments. Second, a scientific revolution within ecology questioned assumptions about nature's passivity and stability. We discuss these in turn in the following sections.

The Indigenous environment

By the 1980s, political activism, centred on the recognition and rights of indigenous Australians, had sufficiently provoked Australian modernity that the legitimacy of the doctrine of *terra nullius* was under wide challenge. This challenge was written into law with the High Court's recognition of 'native title' in 1992. Tensions between advocacy for wilderness and recognition of indigenous cultural landscapes were evident as early as the iconic 1982 environmentalist campaign that 'saved' the Franklin and Gordon Rivers in Tasmania from damming (Head 2000). For many environmentalists, these tensions were resolved as indigenous cultural histories were adopted as examples of sustainable environmental management to be emulated by modern society. Colonial representations of Australia's indigenous peoples as primitive cousins whose lives were shackled to nature are in the process of being displaced in environmental discourses by 'postcolonial' representations of them as sophisticated environmental stewards whose spiritual ethos of environmental interconnectedness is fully consistent with scientific understanding of ecological holism and complex systems (Wensing 2014).

Not all environmentalists allow that indigenous cultures deserve to be cast with nature. For some staunch and influential defenders of nonhuman wilderness, such as William Lines (2006, p. 243–244), the idea that indigenous cultural values are part of a suite of natural values reflects an insidious fusion of environmentalism and humanism that has weakened the cause of nature conservation in Australia. For Lines (2006, p. 354), environmentalists 'trapped in wishful thinking about the wisdom of the elders … are incapable of recognising the truth: there are no models, no templates for living sustainably on this continent'. Yet, despite Lines and others who long for a frontier with which to cleave the realms of humanity and nature, Indigenous people now commonly feature in environmental narratives of precolonial sustainability. Reflecting this, the turn of the twenty-first century saw the idea that the 'natural areas of Australia are cultural landscapes' embedded in environmental management textbooks (e.g. Worboys *et al.* 2001, p. 28).

The integration of indigenous cultural environments within environmentalist accounts of a wild frontier has moved in parallel with the integration of many contemporary indigenous communities within modern environmental management regimes in Australia (Howitt and Suchet-Pearson 2006). Propelled by native title arrangements, co-management between modern land managers and indigenous communities has become common. Ross *et al.* present Australia 'as an international pioneer in co-management of terrestrial protected areas and the declaration and management of Indigenous protected areas' (2009, p. 242). Discrete indigenous practices, such as burning the landscape and cultivating 'bush tucker' (native food), are being reframed as rational and efficient ways of engaging with Australian environments. More generally, the roles and insights of Aboriginal peoples as traditional custodians of 'country' have begun to be incorporated within modern institutions as a form of credible traditional ecological knowledge (Hill *et al.* 2011; Weir 2012).

Emphasis on co-management has, however, exposed the incommensurability of modern and indigenous environmental ontologies (Suchet 2002; Trigger *et al.* 2008; Verran 2009). This difference manifests in the way in which modern distinctions between scientific management and moral experience of nature have no purchase on indigenous worlds. Attempts to sort objects from subjects, facts from affects run aground in indigenous environmental ontologies in which the activities of description and prescription are inseparable. However, this lack of articulation between indigenous and modern ontological practices also manifests in fundamental differences between the environmental space/times to which they give rise.

The linear telos of an advancing modernity whereby the past is irrevocably lost and the future waits to be claimed makes little sense in Australian indigenous environmental ontologies in which ancestral presence is material in a protracted present, one without the constraint of an imminent future (Rose 2004; see also Crabtree 2013). Despite the fact that British invasion initiated a still-active history of slow violence against indigenous worlds, this event appears to be much more pivotal in modern environmental ontologies than it does in the topological temporality of indigenous environments. For Australian modernity, 1788 acts as an ontological ground zero, a starting of the clock of Western history in this continent (Head 2012). This dualistic pivot first held civilised order apart from primordial incoherence and later underpinned efforts to hold true, wild nature from the despoliations of the modern world. In indigenous ontologies, however, invasion does not appear to have a discrete before and after, as the protracted present encompasses invasion and its aftermath within an ancestral presence (Crabtree 2013). This is less a moment than an enduring tear in the fabric of past, present and future: a devastating rending, but one far from final for many Indigenous peoples. It is easy for modern minds to conclude that indigenous environmental ontologies represent some kind of collapse of time into a static present. But for all the modern emphasis on progress as an urgent bid for freedom from the past, indigenous environmental ontologies emerge as much more dynamic than dualistic modern ontologies that seek fixed spatial and temporal reference points. As Rose (2004) explains, the indigenous present is a temporality held open by human-other-than-human practice. This is a space in time that is the result of active systems of spatial and temporal organisation. This is a topological gathering and knotting together of human and nonhuman possibilities, embedded in unfolding worlds of practice.

The dynamism of indigenous environmental ontologies has been emphasised over the past decade through their encounter with a static frontier between natives and invaders in many environmental management regimes. Modern management practices have placed heavy emphasis on the need to control and even eliminate 'introduced', 'exotic' nonhumans in the space/time of authentic Australian nature, routinely conflating the categories of nonnative and invasive species (Head 2012). The fervour with which many modern Australians pursue this war on what government has dubbed 'the weed menace' invites inquiry into how it carries desire for redemption in a colonial society built upon human acts of invasion (Lien and Davison 2010). This fervour is in contrast to the fluid, contextual and adaptive way in which many Indigenous Australian communities have accommodated nonnative species within their stories of becoming and the environmental practices these stories animate (Trigger 2008; Martin and Trigger 2015). Countenancing no absolute outside from which species could be introduced, many Indigenous communities appear responsive in negotiating and moving with environmental change in ways that maintain continuity within ancestral horizons. We next take up the question of how environmental change is encountered within modern ontologies.

The mobile environment

The second way in which the modern frontier between nature and society has been destabilised in Australia in recent decades relates to the emergence of nonequilibrium ecological theory. Challenging the previously dominant scientific assumption that evolutionary systems head towards stasis, the 'new ecology' has unsettled ideas of equilibrium, harmony and balance in ecological processes by emphasising flux, stochastic change, novelty and opportunism (Botkin 1990; Low 2002). This theoretical innovation has been coproduced with intensifying material dynamics of planetary innovation linked to globalised sociotechnical practices. That is, the new ecology is both a reorientation of modern knowledge of living systems and a reflection of modern human implication in the production of novel ecological realities. This twofold opening toward environmental flux in modern societies underpins growing interest in the early years of the twenty-first century in the resilience of coupled socioecological systems (Folke 2006; Hobbs *et al.* 2014).

In Australia, growing acknowledgement of the role of flux and novelty in ecological processes has widened the focus of environmental management well beyond the protection of 'natural' environments. Rather than the distant frontier of wilderness, there has been a notable increase in professional and popular interest over the past 30 years in nonhuman dimensions of the urban and suburban environments in which most Australians live (Davison and Ridder 2006; Byrne *et al.* 2014). While part of this interest has taken the form of attempts to maintain enclaves of true nature in the midst of the artificiality of the city, more recently it has seen conceptual boundaries between nature and society softened and the city rethought as a socioecological whole (Davison and Kirkpatrick 2014).

The new ecology has also challenged the idea that environmental managers should seek to hold Australian environments in a steady state defined by a pre-colonial baseline state of nature. This shift in thinking has resulted in increased emphasis on adaptive forms of management aimed at maintaining capacities, such as resilience or ecosystem services, rather than fixed states (Allen and Stankey 2009). Thus, the federal government's 1996 National Biodiversity Strategy recognised that biodiversity is 'not static, but constantly changing' (Commonwealth of Australia 1996, p. 6).

Implementing adaptive approaches to environmental management that do not rely on a baseline state of nature has proven difficult, as evident in Australia's current Biodiversity Conservation Strategy, 2010–2030 (Commonwealth of Australia 2010). This Strategy reaffirms the recognition in the 1996 Strategy that biodiversity is 'constantly changing', particularly in the context of 'unpredictable interactions between climate change and other factors that cause stress to ecosystems' (Commonwealth of Australia 2010, p. 29). However, it also asserts that biodiversity is 'best conserved by protecting existing natural habitats', placing strong emphasis on the need to combat invasive species (Commonwealth of Australia 2010, p. 16). Tensions implicit in the Strategy between preexisting and new natures surfaced during the consultation process. Consider the suggestion in the 2009 consultation draft 'that ecological systems change naturally, in unpredictable and sometimes unexpected ways … [and thus] we must embrace and work with natural ecological variability, rather than attempting to control or reduce such variability' (National Biodiversity Strategy Review Task Group 2009, p. 20). The Draft Strategy provoked a 'letter of concern' from 90 of Australia's senior nature conservation scientists (Arthington and Nevill 2009). This group expressed alarm at what they saw as a retreat from an existing policy commitment to ensure a 'comprehensive, representative and adequate system of ecologically viable protected areas' (Arthington and Nevill 2009, p. 78). The absence in the final strategy of the strong acknowledgement of ecological variability of the Draft Strategy can presumably be attributed to this and similar feedback.

Recognition of ecology as a dynamic process rather than as fixed states is difficult in Australia because it destabilises the space/time of Australian modernity. The Australia Day Address by 2002 Australian of the Year and environmental scientist Tim Flannery reveals what is at stake for many who hold on to an Australian nature firmly fixed in time and space:

> Australia – the land, its climate and creatures and plants … [is] the only force ubiquitous and powerful enough to craft a truly Australian people.

> For Australians, the land has a special significance …. Australia … has remained almost unique in its stability. Its biodiversity increased in relative peace and isolation over the eons …. And because of that stability many species became very specialised, confined perhaps to just a few square kilometres, making them vulnerable to future changes.
>
> (2002, n.p.)

For Flannery and many other environmentalists, a uniquely Australian nature anchored in a precolonial order holds out the prospect of authentic Australian-European belonging in the aftermath of invasion. This desire for roots in nature exists in tension with efforts to adaptively manage dynamic socioecological systems. These tensions are being exacerbated by the prospect of wholesale environmental change linked to anthropogenic climate change.

As noted at the outset, the future prospect of a radically foreign Earth is being narrated as an Anthropocene. This is a time on Earth when humanity holds the future in its hands: a global ground zero from which the arrow of time departs again. Stories of the Anthropocene comprehend contemporary earth dynamics through modern ontologies that assert humanity's essential difference and that claim human exemption from the rules that govern the rest of reality. These ontologies take epistemological form as a logic of exaggeration whereby human implication in environmental flux is taken as evidence of the end of nature's autonomy. Earlier environmentalist metaphors about spaceship Earth harden into scientific description of the all-too-real workings of 'The Human Earth'. In the concluding section, we critically examine narratives of the Anthropocene by briefly considering their relevance for our discussion of the ontological practice of the environment in Australia.

The crisis environment

More than anything else, modern narratives of the Anthropocene are borne out of deep anxieties about environmental crisis, an emergency in the present that, if left unattended, calls forth future global catastrophe. This is a crisis in which the fate of humanity is inseparable from the fate of Earth's diverse freight of nonhuman life. A central message of these narratives is that the stability of nature, as a foundation upon which humanity can build and improve, is no longer assured. The globalised practices of twenty-first-century modernity have established mobility as an organising principle of human life (Urry 2011). Dependent on fossil fuels, these conditions have unexpectedly put the whole Earth in motion. Organisms, things, forces and affects: all unmoored on a planet growing strange. According to narratives of the Anthropocene, any future stable basis for human survival, let alone progress, will be founded on conscious human stewardship aimed at maintaining the planetary conditions for life (Steffen *et al.* 2011).

Universalism is a hallmark of stories of the Anthropocene. They are based upon what Rob Nixon (2014, n.p.) calls a 'grand explanatory species story' of human folly and (so the narrator of this story hopes) human redemption. Within these stories, there is little room for the deep and enduring injustices and inequities between humans that have characterised the imperialist, capitalist and technoscientific histories of modern progress. Recent incorporation of the uneven geographies of 'human development' within these narratives only entrench understanding of modern progress as an ineluctable law of human evolution (see, e.g., Steffen *et al.* 2015). 'Slender abstractions' about the human species effectively collapse human difference (Malm and Hornborg 2014, p. 4). Abstractions about earth systems and planetary forces, despite their link to advocacy for biodiversity, also have the potential to collapse environmental difference.

In the context of the Anthropocene, the space/times of Australian environments and their relation to Australian modernity are hard to discern, subordinated as they are to global space and epochal time. The Anthropocene echoes the founding conditions of Australian modernity in which pioneers sought to accommodate themselves in an utterly foreign reality. The pioneer response was to establish a frontier with which to hold primordial nature at bay and behind which an old world was given new expression. This frontier was pushed across the continent (and its archipelago), entwining spatial expansion with the teleological temporality of progress. The figure of nature was relegated to the margins of modern order, being oddly understood as both wild and containable in timeless spaces of protection, and becoming the focus of cultural yearning for an endangered past. This yearning has increasingly encompassed indigenous cultural histories, although not necessarily contemporary indigenous peoples, and the company of nonhuman natives has been sought in suburban gardens, children's stories and gourmet menus.

In the Anthropocene, modern Australians are confronted with a world in which time threatens to erase space and in which a global epoch forecloses local possibility. This is the danger – that a newfound ground of modern Australian belonging in native nature is melting away. Australia's environmental managers are urged to hold on to at least a representative sample of natural environments and are charged with projects of ecological border protection. Although beyond the scope of this discussion, it is worth considering how such ecological securitisation relates to an intense political focus in the early years of the twenty-first century on social border protection from unwanted human arrivals (Lien and Davison 2010).

Talk of the Anthropocene extends the sphere of modern concern beyond familiar registers of 'experiential and historical time' into the expanses of 'climatological time' (Markley 2012), so as to draw ecological futures into the present. Yet, with the thought that human signatures are written on every molecule of the atmosphere, the figure of Australian wilderness becomes even more alluring for being ever more elusive. Planeloads of international tourists add their names to the new atmospheric petition in the process of gazing upon the feted (and fated) difference of antipodean nature. Australian modernity itself is among the most highly carbon-intensive of all contemporary societies. Thus prospective anxieties of untenable but deeply embedded ways of living in the present are now added to the anxieties that have come with the founding of this nation on what is retrospectively becoming an untenable act of invasion.

The story of the Australian Anthropocene is of a continent cut adrift from history and of a society seeking postcolonial roots in a nature that will not stand still. This is a story that masks the prospect that, far from being fully human, contemporary Australian environments are becoming newly wild. At a time when environmental managers are told they have nothing less than the conditions of life, including human life, in their hands, environments are growing less intelligible and predictable.

In this chapter we have sought to retell the modern history of Australian environments through an interest in the modern ontological practices that work to deny the seamless existence of humanity within wholes that exceed them, seeking the universal reason that comes with distance. Despite growing material evidence of the incoherence of modern distinctions between nature and humanity, the ontological work of separation continues in the form of stories of an Anthropocene in which difference is erased in a global time of human dominance. Our aim has ultimately been to call this work of separation into question as well as to catch a glimpse of the other times and spaces that may run in and through Australian environments.

References

ABS. 2003. *Australian Social Trends*, Cat. No. 4102.0. Canberra, Australia: Australian Bureau of Statistics.

Allen, C. and Stankey, G.H. 2009. *Adaptive Environmental Management: A Practitioner's Guide*. Dordrecht, The Netherlands and Melbourne, Australia: Springer and CSIRO.

Anderson, K. 2008. *Race and the Crisis of Humanism*. London, England: Routledge.

Arthington, A.H. and Nevill, J. 2009. Australia's biodiversity conservation strategy 2010–2020: Scientists' letter of concern. *Ecological Management and Restoration*, 10(2), 78–83.

Australian Government. n.d. Australian National Anthem. *It's an Honour: Australia Celebrating Australians*. Accessed 1 July 2015. Available at http://www.nla.gov.au/apps/cdview/?pi=nla.mus-an24220024.

Bernstein, R. 1983. *Beyond Objectivism and Relativism: Science, Hermeneutics, and Praxis*. Philadelphia, PA: University of Pennsylvania Press.

Botkin, D.B. 1990. *Discordant Harmonies: A New Ecology for the Twenty-First Century*. New York, NY: Oxford University Press.

Byrne, J., Snipe, N. and Dodson, J. (eds). 2014. *Australian Environmental Planning: Challenges and Future Directions*. Abingdon, England: Routledge.

Carter, P. 1987. *The Road to Botany Bay: An Essay in Spatial History*. London, England: Faber and Faber.

Castree, N. 2014. The Anthropocene and geography I: The back story. *Geography Compass*, 8(7), 436–449.

Clark, N. 2011. *Inhuman Nature: Sociable Life on a Dynamic Planet*. London, England: Sage.

Commonwealth of Australia. 1996. *National Strategy for the Conservation of Australia's Biological Diversity*. Canberra, Australia: Commonwealth of Australia.

Commonwealth of Australia. 2010. *Australia's Biodiversity Conservation Strategy, 2010–2030*. Canberra, Australia: Natural Resource Management Ministerial Council.

Crabtree, L. 2013. Decolonising property: Exploring ethics, land, and time, through housing interventions in contemporary Australia. *Environment and Planning D: Society and Space*, 31(1), 99–115.

Crutzen, P.J. and Stoermer, E.F. 2000. The Anthropocene. *Global Change Newsletter*, 41, 17–18.

Davison, A. 2001. *Technology and the Contested Meanings of Sustainability*. Albany, NY: State University of New York Press.

Davison, A. 2005. Australian suburban imaginaries of nature: Towards a prospective history. *Australian Humanities Review*, 37(December). Accessed 6 September 2015. Available at http://www.australianhumanitiesreview.org/archive/Issue-December-2005/davison.html.

Davison, A. 2015. Beyond the mirrored horizon: Modern ontology and amodern possibilities in the Anthropocene. *Geographical Research*, 53(3), 298–305.

Davison, A. and Kirkpatrick, J.B. 2014. Re-inventing the urban forest: The rise of Australian arboriculture. *Urban Policy and Research*, 32(2), 145–162.

Davison, A. and Ridder, B. 2006. Turbulent times for urban nature: Conserving and re-inventing nature in Australian cities. *Australian Zoologist*, 33(3), 306–314.

Davison, G. 1995. Australia – the first suburban nation. *Journal of Urban History*, 22(1), 40–74.

Devlin-Glass, F. 1994. 'Mythologising spaces': Representing the city in Australian literature, in *Suburban Dreaming: An Interdisciplinary Approach to Australian Cities*, edited by L.C. Johnson. Melbourne, Australia: Deakin University Press, 160–180.

Flannery, T. 2002. *The Day, The Land, The People*, Australia Day Address 2002, Sydney Conservatorium of Music, 23 January. Accessed 6 September 2015. Available at http://cdn.australiaday.com.au.s3-ap-southeast-2.amazonaws.com/wp-content/uploads/2014/01/21090055/Read-the-2002-Australia-Day-Address.pdf.

Folke, C. 2006. Resilience: The emergence of a perspective for social-ecological systems analysis. *Global Environmental Change*, 16, 253–267.

Franklin, A. 2006. *Animal Nation: The True Story of Animals and Australia*. Sydney, Australia: University of New South Wales Press.

Gascoigne, J. 2002. *The Enlightenment and the Origins of European Australia*. Cambridge, England: Cambridge University Press.

Head, L. 2000. *Second Nature: The History and Implications of Australia as Aboriginal Landscape*. Syracuse, NY: Syracuse University Press.

Head, L. 2012. Decentring 1788: Beyond biotic nativeness. *Geographical Research*, 50(2), 166–178.

Heidegger, M. 1977. *The Question Concerning Technology and Other Essays*, translated and edited by W. Lovitt. New York, NY: Harper & Row.

Hill, R., Walsh, F., Davies, J. and Sandford, M. 2011. *Our Country Our Way: Guidelines for Australian Indigenous Protected Area Management Plans*. Cairns, Australia: CSIRO Ecosystem Sciences and Australian Government Department of Sustainability, Water, Environment, Population and Communities.

Hobbs, R.J., Higgs, E., Hall, C.M., Bridgewater, P., Chapin, F.S., Ellis, E.C., *et al.* 2014. Managing the whole landscape: Historical, hybrid, and novel ecosystems. *Frontiers of Ecology and Environment*, 12(10), 557–564.

Howitt, R. 2001. Frontiers, borders, edges: Liminal challenges to the hegemony of exclusion. *Australian Geographical Studies*, 39(2), 333–345.

Howitt, R. and Suchet-Pearson, S. 2006. Rethinking the building blocks: Ontological pluralism and the idea of 'management'. *Geografiska Annaler B*, 88(3), 323–335.

Johnson, E., Morehouse, H., Dalby, S., Lehman, J., Nelson, S. Rowan, R., *et al.* 2014. After the Anthropocene: Politics and geographic inquiry for a new epoch. *Progress in Human Geography*, 38(3), 439–456.

Latour, B. 1993. *We have Never Been Modern*, translated by C. Porter. Cambridge, MA: Harvard University Press.

Latour, B. 2004. *Politics of Nature: How to Bring the Sciences into Democracy*, translated by C. Porter. Cambridge, MA: Harvard University Press.

Lien, M. and Davison, A. 2010. Roots, rupture and remembrance: Contesting here and there, then and now in the Tasmanian lives of Monterey Pine. *Journal of Material Culture*, 15(2), 233–253.

Lines, W.J. 1991. *Taming the Great South Land: A History of the Conquest of Nature in Australia*. Sydney, Australia: Allen & Unwin.

Lines, W.J. 2006. *Patriots: Defending Australia's Natural Heritage*. Brisbane, Australia: University of Queensland Press.

Low, T. 2002. *The New Nature: Winners and Losers in Wild Australia*. Melbourne, Australia: Penguin.

Malm, A. and Hornborg, A. 2014. The geology of mankind? A critique of the Anthropocene narrative. *The Anthropocene Review*, 1, 62–69.

Markley, R. 2012. Time: Time, history and sustainability, in *Telemorphosis: Theory in the Era of Climate Change, Vol. 1*, edited by T. Cohen. Ann Arbor, MI: Open Humanities Press, 43–64.

Martin, R.J. and Trigger, D. 2015. Negotiating belonging: Plants, people, and indigeneity in northern Australia. *Journal of the Royal Anthropological Institute*, 21, 276–295.

Morton, T. 2013. *Hyperobjects: Philosophy and Ecology after the End of the World*. Minneapolis, MN: University of Minnesota Press.

National Biodiversity Strategy Review Task Group. 2009. *Australia's Biodiversity Conservation Strategy 2010–2020, Consultation Draft*. Canberra, Australia: Australian Government, Department of the Environment, Water, Heritage and the Arts.

Nixon, R. 2011. *Slow Violence and the Environmentalism of the Poor*. Cambridge, MA: Harvard University Press.

Nixon, R. 2014. The Anthropocene: The Promise and Pitfalls of an Epochal Idea. *Edgeeffects*, 6 November. Accessed 6 September 2015. Available at http://edgeeffects. net/anthropocene-promise-and-pitfalls/.

Rose, D. 1997. The year zero and the North Australian frontier, in *Tracking Knowledge in North Australian Landscapes: Studies in Indigenous and Settler Ecological Knowledge Systems*, edited by D. Rose and A. Clarke. Canberra, Australia: North Australia Research Unit, Australian National University, 19–36.

Rose, D. 2004. *Report from a Wild Country: Ethics for Decolonisation*. Sydney, Australia: University of New South Wales Press.

Rose, D. and Clarke, A. (eds). 1997. *Tracking Knowledge in North Australian Landscapes: Studies in Indigenous and Settler Ecological Knowledge Systems*. Darwin and Canberra, Australia: North Australia Research Unit, Australian National University.

Ross, H., Grant, C., Robinson, C.J., Izurieta, A., Smyth, D. and Rist, P. 2009. Co-management and Indigenous protected areas in Australia: Achievements and ways forward. *Australasian Journal of Environmental Management*, 16(4), 242–252.

Steffen, W., Broadgate, W., Deutsch, L., Gaffney, O. and Ludwig, C. 2015. The trajectory of the Anthropocene: The great acceleration. *The Anthropocene Review*, 2(1), 81–98.

Steffen, W., Persson, A., Deutsch, L., Zalasiewicz, J., Williams, M., Richardson, K., *et al.* 2011. The Anthropocene: From global change to planetary stewardship. *Ambio*, 40, 739–761.

Suchet, S. 2002. 'Totally Wild'? Colonising discourses, indigenous knowledges and managing wildlife. *Australian Geographer*, 33(2), 141–157.

Trigger, D. 2008. Indigeneity, ferality and what 'belongs' in the Australian bush: Aboriginal responses to 'introduced' animals and plants in a settler-descendant society. *Journal of the Royal Anthropological Institute,* 14(3), 628–646.

Trigger, D., Mulcock, J., Gaynor, A. and Toussaint, Y. 2008. Ecological Restoration, cultural preferences and the negotiation of 'nativeness' in Australia. *Geoforum,* 39, 1273–1283.

Urry, J. 2011. *Climate Change and Society.* Cambridge, England: Polity Press.

Verran, H. 2009. Natural resource management's 'nature' and its politics. *Communication, Politics & Culture,* 42(1), 3–18.

Weir, J.K. (ed.). 2012. *Country, Native Title and Ecology.* Canberra, Australia: Australian National University E-Press.

Wensing, E. 2014. Aboriginal and Torres Strait Islander peoples' relationships to 'Country', in *Australian Environmental Planning: Challenges and Future Prospects,* edited by J. Byrne, N. Snipe and J. Dodson. Abingdon, England: Routledge, 9–20.

Worboys, G., Lockwood, M. and De Lacy, T. 2001. *Protected Area Management: Principles and Practices,* 1st ed. South Melbourne, Australia: Oxford University Press.

3 Landscape, temporality and responsibility

Making conceptual connections through alien invasive species

Gunhild Setten

Introduction

There is a hardy and beautiful shrub situated among herbs and grasses 3 metres off the driveway and about 20 metres away from my grandparents' house. The shrub is untidy as it was never granted access to the garden and kept under control. It is a Japanese rose, a *Rosa rugosa*. This particular rose lives in the north of Norway, on the edge between solid ground and the North Atlantic Ocean, and has over the years become a sort of friend greeting me when returning to my grandparents' house in the summer. It has taken on an emotional value, becoming a symbol of holidays long gone and the loss of close family, yet not ceasing to be a *plant*, a living non-human being in a landscape, feeding on soil and sun.

In 2012 the Japanese rose was blacklisted in Norway as an alien invasive species in the category 'very high risk'; that is, the rose is considered to be a species threatening to damage biological diversity and disperse to the natural environment (Gederaas *et al.* 2012). This has not changed my relationship with this particular specimen, yet the Japanese rose has, in generic terms, taken on a totally different meaning within environmental management and policy making. Its status has shifted from an object of pride to a pest, reflecting a fundamental shift in value judgements. This is, as I write, a legal fact. A new regulation under the Norwegian Nature Diversity Act (2009) was released on 19 June 2015 (to take effect 1 January 2016), prohibiting the import and sale of invasive alien species categorised as high ecological risk in the 2012 Norwegian Blacklist (Gederaas *et al.* 2012). The Japanese Rose is hence legally defined as unwanted in gardens and parks as well as beyond any fence. *My* rose is thus cast as a pest by the environmental authorities, 'a problem to be eradicated, a species to be regulated, an impediment to mustering and a non-native' (Head *et al.* 2015, p. 410). In short, it is now an adversary, a symbol of a degrading and damaged Nature for which we humans need to take responsibility.

In this chapter, and with a particular focus on Norway, I use the example of (invasive) alien plant species in order to conceptualise and problematise connections between landscape, temporality and responsibility. Inspired by Bruno Latour's notion of the 'multinatural', and related thinking in geography (Lorimer 2012; see, e.g., Atchison and Head 2013; Robbins and Moore 2013; Head *et al.* 2015;

Qvenild and Setten Forthcoming), I argue that any ecology is a process where values are mobilised and contested. The processual nature of such an ecology challenges our thinking about landscape, temporality and responsibility. I therefore want to bring the 'multinatural' into explicit conversation, firstly, with landscape; that is, the physical and symbolic spaces where shifting human–plant relationships unfold. Second, with time; that is, nature's fluidity invites a temporal element. And finally, with responsibility; that is, because 'our values contribute to our pest problems' (Low 2001, p. 36), only humans can mobilise a sense of responsibility, or not, for halting the spread of alien species.

This chapter represents mainly a conceptual conversation and is structured as follows: I start by offering a short outline of the alien invasive species debate, both managerially and intellectually, which actualises notions of, and connections between, landscape, temporality and responsibility. I move on by bringing these three related strands of thought and action emanating from the species debate into conversation, before returning to the debate more explicitly towards the end.

Writing primarily from a landscape perspective, that is, a perspective where landscape is mobilised both physically and symbolically, I aim to make explicit a tension between past, present and future in species management, which becomes visible when using responsibility and landscape as lenses through which to examine this tension. I particularly focus on questions concerning responsibility and shifting discursive and material contexts, actualising temporality for how responsibility is conceptualised and manifested in landscapes. One of my main concerns is whether 'landscape' provides an opportunity to inform these questions. I complete the chapter by suggesting what I think are opportunities for landscape research when problematising and taking responsibility for what appears to be an open, yet uncertain, future.

Alien invasive species: from rational to relational thinking

The mobility of non-human species, often aided by humans, is by now established as one of the most serious threats to biodiversity globally (Ministry of the Environment 2007; Gederaas *et al.* 2012). Consequently, all nations have a responsibility to protect their biodiversity against loss and destruction. By committing to the 1992 UN Convention on Biological Diversity (CBD), a nation is legally obliged to take measures in order to identify potentially ecologically harmful species as well as identifying species at risk of extinction. The blacklisting of harmful species (e.g. Gederaas *et al.* 2012) and the more common redlisting of endangered species (e.g. Kålås *et al.* 2010) have become key tools in order to meet the obligations of the CBD in Norway, as well as in other countries. The intention with these listings is to enable informed decisions over which species to conserve or protect and which to eradicate. These lists contain species categorised as endangered and consequently native, as alien or as invasive alien. A native is a species 'occurring within its natural range (past or present)' (IUCN Council 2000, n.p.), while an alien is a species which manages to reproduce and survive outside its natural range and which has reached new locations often with the help of humans. An alien species threatening biodiversity is categorised as an alien *invasive* species. As at 2012, 2,320 alien species had been recorded in Norway, of which 216 were categorised as having 'severe impact' or 'high impact' on native nature (Gederaas *et al.* 2012).

Neither 'alienness' nor 'nativeness' can be observed per se, hence distinguishing aliens from natives is a value-based demarcation of spatial belonging of any species at a particular time (Qvenild 2014; see also Warren 2007). The CBD provides a spatial definition of alien and native, but a temporal designation is lacking, leaving it up to policymakers and scientists to demarcate a temporal reference point, that is, a baseline or a 'threshold of nativeness' (Head and Muir 2004, p. 202; see also Smout 2003; Head 2012; Robbins and Moore 2013; Qvenild and Setten Forthcoming). In Norway, the year 1800 is set as year zero, that is, species introduced to Norway after 1800 are cast as aliens. Interestingly, no clear explanation has been given for why 1800 is a point of reference. Gederaas *et al.* only state that: 'The year 1800 is used as the historical time limit for risk assessments, and assessments of future risks are limited to species that have the potential to become established in the next 50 years and that could pose an ecological impact in Norway during that time' (2012, pp. 7–8). It appears, then, that the year 1800 is chosen primarily 'for regulatory purposes and for providing the authorities with a possibility to distinguish what is assumed to belong in Norwegian nature from that which is not' (Qvenild 2014, p. 184), including avoiding species ending up on both lists.

The strategy of Norwegian environmental authorities to combat alien invasive species is not unique (see Smout 2003; Warren 2007; Head 2012, for related examples). Listing, systematising and hierarchising Nature are well-established international practices developed to exercise control over species and habitats, yet, according to some, 'not to the extent we would like' (Low 2001, p. 36). Low argues that:

> Values that developed over hundreds of years ago strongly influence us today. Those that contribute to our pest problems [that is, alien invasions] include love of mobility, freedom, speed, diversity, progress, familiarity, and a mechanistic view of nature.
>
> (2001, p. 37)

According to Low (2001), these values hamper our ability to exercise control, both practically and intellectually. The paradoxical nature of current species politics is thus closely related to what in many ecological quarters is held as a fact, that 'we want a world in which people are as free as possible to travel and to exchange goods and ideas. ... But at the same time, we *need* a world in which most other living things stay put' (Low 2001, p. 41, emphasis in original). Such a 'fact' has led Qvenild and Setten (Forthcoming) to argue that 'a systematised and hierarchized representation of nature forces a critical discussion of practices resulting from the formal establishment of degraded nature'. The alien species debate and its associated practices, which are ultimately set within larger environmental debates, involve material and discursive attempts to mobilise, fix and purify landscapes according to what appears to be random temporal species rhetoric. Such practices have proved to be of critical political and practical consequence. We need, then, to explore how temporality matters in shaping the relationship between landscape and responsibility, in the past, present and future. The point is not that the relationship between landscape and responsibility has a temporal element, but rather how the relationship is set to work when *a future* is up for grabs. In the literature, this has so far not been explicitly problematised.

I offer a problematisation and an illustration, taking my cue from Qvenild (2012, 2014), who in a Norwegian context has pioneered work on how species categories given in the redlists and blacklists are set to work in environmental conflicts. She has demonstrated the critical role of temporality for species politics and rhetoric, most explicitly in her study of the redevelopment of the former main Norwegian airport at Fornebu in Oslo. At Fornebu, temporal thresholds set in the past were mobilised by clashing interests in order to justify – or, alternatively, to contest – the choice of a future landscape (Qvenild 2012, 2014). Within the Fornebu context, responsibility is legally defined, particularly through the CBD, the Norwegian blacklists and redlists (Kålås *et al.* 2010; Gederaas *et al.* 2012) and the Norwegian Nature Diversity Act. This means they all problematise how plants cause harm through their agency, as well as the degree of human control over such agency. In consequence, they identify a threat to life in general that needs to be dealt with before it reaches a point of irreversibility (Anderson 2010, p. 789). Making the future present, then, becomes a key way to enact such a threat. The debate about species is therefore also a debate about moralities, judgements and, not least, deep suspicion directed towards both humans and, in the current case, plants. Judicial tools, such as the Norwegian Nature Diversity Act, confirm the constant challenge of disorder represented by the movement of species by humans.

Qvenild's work, to which I will return, is only one of a number of works supporting Robbins and Moore's (2013, p. 4) claim that debate about alien invasive species 'carries complex interpretive baggage'. This has led Head *et al.* (2015, p. 399, see also Lorimer 2012), among others, to argue more broadly that 'the need for diverse scholarship on our relationships with plants has never been greater'. This need reflects an intellectual 'relational turn' where focus is increasingly placed on what plants do rather than what they are (Lorimer 2012), that is, plants are no longer pure and timeless objects (best) 'removed from Society' (Lorimer 2012, p. 594) and held in restraint by humans. The agency of plants has increasingly come to the fore and been demonstrated within diverse contexts, including trees (Jones and Cloke 2002), weeds (Head 2014) and gardens (Qvenild *et al.* 2014). In short, 'multinatural' worlds, 'characterized by lively processes and impure forms, co-existing in inhabited landscapes', are challenging 'the privileged place of the human subject in accounts of environmental change' (Lorimer 2012, p. 595. See Davison and Williams, this volume.). Human–nature relations are increasingly destabilised, with critical implications for how we are able to think about human and non-human life in general, and future human-non-human life in particular. Against this background, I turn now to discuss implications for landscape, temporality and responsibility as an effect of such a destabilisation.

Landscape, temporality and responsibility – a conceptual conversation

Given the vast literature on 'landscape', here I subscribe to a landscape discourse concerned with the meaning and production of the material landscape (e.g. Setten 2004). This is a discourse holding that landscape is fundamentally political: the contested processes through which landscapes are shaped are key to understanding how landscapes are both means and ends in shaping social relations. *Because* landscape is political, it is both a material fact and a representational social construct. Landscapes, then, are sites of contention, claims and contestations, conceptually, politically, emotionally and sometimes literally (see Warren 2007). In effect, they are struggled over and are the means of struggle (Mitchell 2012); that is, landscapes are mobilised to regulate and control behaviour.

Social struggles not only shape landscapes but crucially also involve attempts to naturalise them, making them seem inevitable, ordinary and even necessary (Setten and Brown 2009). Social struggles are also attempts to resist such naturalisation. So, landscapes work to (re)produce certain identities and ways of life and become a spatial configuration of the legitimacy and moral authority of particular people. In short, the landscape is neither untroubled nor untroubling.

Together with Katrina M. Brown, I have argued that strands in current critical landscape research – mainly in geography – are dealing with a notion of landscape where landscape plays a prominent analytical role in shedding light on its troubling nature for social relations. These strands are concerned with policy, justice, morality, labour, class and production, race and memory, everyday struggle and belonging (Setten and Brown 2013). They all allude to responsibility, and they are both empirically and analytically concerned with how landscape works when we 'take action' – or not – against social and environmental injustices or the movement of species, for example. Responsibility, then, is on the whole taken for granted and appears to be a necessary condition for a discussion about injustices or alienation. From a landscape perspective, there are hence good reasons for reflecting on responsibility more explicitly.

According to Staeheli (2013, p. 523), 'responsibility talk seems pervasive'. We *take* responsibility or *act* responsibly every day, in the simplest sense meaning that we, in Doreen Massey's (2004) terms, 'do something about it', whatever that 'something' or 'it' refers to. Responsibility is often conceptualised, either morally or legally, as taking care of each other, sharing with each other and being kind to each other – in short, meeting the needs and being attuned to the welfare of others in a community and acting accordingly (Staeheli 2013). Responsibility is hence related to active citizenship, and 'acting responsibly is ... internalised by "good" citizens' (Staeheli 2013, p. 524). By implication, *not* taking responsibility or acting irresponsibly becomes a normative break with expectations which are either morally or legally enforced by the self, other individuals, groups or institutions.

In our daily lives we are all experts when it comes to that more or less intuitive sense of responsibility, that is, knowing when and how to act, or not, in particular circumstances. The exact social meaning of responsibility is thus contextual. Academically there is a vast literature addressing responsibility, but it appears to be an underlying, taken-for-granted notion surfacing in discussions about public participation, moral and ethical issues or citizenship, for example. Responsibility frequently seems to come under a different name, both within and beyond what we might call the 'responsibility literature'. The notion of responsibility is hence wide and challenging. It appears reasonable to claim, however, that 'responsibility talk' within academia in general is mainly part of a radical, activist language entangled with political commitment. In geography, if there is such a thing as a geography of responsibility, a political commitment is reflected in the works of David M. Smith (2000), who must be seen as the instigator of what has been termed a 'moral turn' in geography. Smith's work has much bearing on the complex spatiality of responsibility and caring. By urging geographers to consider the *scope* of responsibility and morality, Smith (2000) has challenged us to critically consider our capabilities to 'do something' over distances and beyond our control. It is, however, Doreen Massey's journal article 'Geographies of responsibilities' (2004) on which many geographers are leaning when seeing the relevance of being explicitly sensitive to how responsibility works in shaping identities across places and scales. Massey, in many respects echoing Smith, argues for a geography of responsibility which is relational, embodied and spatially extended: 'Responsibility ... derives from those relations through which identity is constructed' (2004, pp. 9–10). So, to perhaps oversimplify, by punctuating place as a coherent category and, by implication, understanding place as a relational achievement, we are equipped to act responsibly across scales, according to Massey.

There is much to commend in Massey's attempt to formulate a relational geography of responsibility. Massey's arguments expressed in her 2006 journal paper 'Landscape as provocation' are, however, more suitable for this occasion. This might come across as slightly ironic as she talks neither about responsibility as such nor about landscape in a very enthusiastic way. She does raise the question of temporality for landscape, however. Fundamental to Massey's occasional concern with landscape is scepticism towards it. This scepticism is based in what Massey (2006) finds to be its preoccupation with settledness, localism and dwelling. I agree with her argument against the tendency within much landscape research to fixate and confine the social to local, material landscapes through ideas about belonging. Her well-known theorisation of place is an argument for open, ongoing and fluid relations, that is, 'place as meeting place rather than as always already coherent' (Massey 2006, p. 34). In short, she argues for an appreciation of place which does not entail parochialism, by insisting on a reconceptualisation of 'the local', yet hanging on to 'a genuine appreciation of the specificity of local areas' (Massey 2006, p. 34).

Massey has called upon landscape researchers to take up the challenges of globally open landscapes and to be more attentive to various connections and disconnections and mobilisations of representational and material landscapes across space and scale. This is, however, not only a spatial or scalar concern. Global openness has temporal consequences, and for Massey it entails a much more explicit and critical perspective on temporality in general, and history/past in particular, on the part of landscape researchers. Again, according to Massey (2006), the compulsion to read the landscape through history tends to reinforce a local, inward focus which narrows and obscures the *spatial* depth of landscape. Massey then is arguing for a perspective which denies landscape its smoothing effect, 'its subtle operation of reconciliation. The conventional continuity of landscape ... is punctuated by a multiplicity of stories' (Massey 2011, n.p.), which contests any form of naturalization. This is not new to landscape researchers. Other works have punctuated what might be seen as culturally and temporally monolithic landscapes and demonstrated how landscapes are spatially extended and contested *because* they are temporal (e.g. Germundsson 2004; Mitchell 2012; Widgren 2012).

I agree with Massey in her insistence that a relational spatiality does not make sense without seeing how past and present become mobilised. My point is that the future is equally important as the past and the present. This is important for the alien species debate as well as for landscape research – landscape being so powerful at 'holding' history, not to say 'holding on' to history. To borrow Barbara Bender's words, '*Landscape is time materialized*' or, more correctly, as she also points out, '*Landscape is time materializing*: landscapes, like time, never stand still' (2002, p. 103, emphasis in original). Landscape and time are inherently processual and in a so-called constant state of becoming. This has become a somewhat canonised statement over the last decades. Even so, and in much landscape research claiming to be processual, there is a sense that landscape *is* history, mainly because the social becomes solidified in and through the landscape. This is not to say that the past is not important, quite the opposite. The problem, however, is that the future is taken for granted when the past is so powerfully mobilised through and in the landscape in order to claim a present. This is also not to say that the future is absent in geography or related fields. According to Anderson (2010, p. 778): 'On the contrary, we find hints of the complicated interrelations between past, present and future across a range of work'. Yet, although there are notable exceptions, geographers rarely engage *explicitly* with the future as a category (Anderson 2010; see Westholm 2012; Brown 2015; Harvey 2015). This is a challenge for landscape research and for the conceptualisation of landscape. We need to understand more explicitly how temporality works and matters in shaping the relationship between landscape and responsibility by thinking about how the boundaries of responsibility are extended in time by also mobilising the future landscape. This means understanding how the future is actually present in and through landscape by way of how it is *made* present (e.g. Adam and Groves 2007). This is another challenge if we accept that landscape has this immense ability to smooth and naturalise peoples' stakes and responsibilities in it and towards it. Landscapes are very good at holding on to versions of what has happened, but what about what has not happened, or only happening to a certain degree – for example, the expected disorder resulting from the movement of species? I now return to Qvenild's work at Fornebu (2012, 2014).

'Sustainable Fornebu'

'Sustainable Fornebu' is the title of one of Norway's largest and most ambitious redevelopment projects (Qvenild 2012, 2014). The project was started in the late 1990s, a result of the closing of the main Norwegian airport in Oslo, and is due to be 'finalised' by 2020. The former airport area, located at the Fornebu peninsula, is a biodiversity hotspot, stated through the establishment of a number of nature reserves, as well as a hotspot for up-market housing and business development close to the city centre of Oslo. This is a match not made in heaven. Statsbygg, leading the redevelopment project on behalf of the Norwegian state, saw it as paramount to take environmental responsibility in developing the area, including protecting birds, insects and plants from the impacts of residents and employees at Fornebu. Species were to be protected mainly through the establishment of buffer zones between the nature reserves and the residential and business areas to be developed (Qvenild 2012, 2014). According to Qvenild (2014, p. 191), when constructing the buffer zones, landscape architects and planners acting on behalf of Statsbygg 'wanted to collect native seeds of, e.g. Blackthorn, hazel and hawthorn for propagation'. However, they encountered a number of obstacles: the natural production of seeds and nuts was too poor; 'the greeneries were inexperienced with how to propagate local seeds'; and there was not enough time for trial and error (Qvenild 2014, p. 191). This resulted in the planting of 'several alien plants with similar qualities' (Qvenild 2014, p. 191), including Blackthorn of Danish origin. Establishing the vegetation in the buffer zones hence quickly became subject to contestation and conflict relating to the use of alien plants. A group of conservationists strongly opposed Statsbygg's ignorance of blacklisted and, by implication, redlisted species, their select view on the nature and qualities of the past landscape on the peninsula, and their rush to establish the buffer zones to the detriment of the production of native bushes and trees. It culminated in the conservationists reporting Statsbygg to the police in 2007 (Qvenild 2012).

The conservationists and Statsbygg were divided over the meaning of responsible behaviour, yet they were joined by spatial and temporal arguments, however differently they portrayed what ought to be 'future Fornebu'. In essence, the landscape, both discursively and materially, was mobilised in a conflict over planning decisions where *a future* is at stake. And when arguing for a certain future, different pasts were mobilised in order to fix both pasts and futures.

Towards conceptual connections

'Sustainable Fornebu' very much reflects current anxieties about damaging and destroying natures that we do not really know the nature of, and 'life without Nature is proving confusing' as Lorimer has aptly pointed out (2012, p. 593). By implication, this and similar projects also reflect current anxieties about making mistakes. To understand how (ir)responsible behaviour occurs in such a context, we need to understand how nature becomes constituted, by whom and by what means. We might ask on which grounds are these plants placed, or alternatively displaced, in conflicts like this.

Conservation of non-human species, both plants and animals, is conditioned by a sense of loss, emotionally and materially, where responsibility for the conservation outcomes is both back- and fore-casted; that is, the future matters as much as the past in any effort to preserve nature. Yet what was planted at Fornebu 10–15 years ago was not only conditioned by the fact that a number of plants used in the buffer zones were not blacklisted at the time. Statsbygg also justified its practices by relying on and taking a certain past for granted (Qvenild 2014), while risking the creation of uncertain future ecologies. Landscape is not considered a dynamic whole, but is being fixed, purified and compartmentalised into buffer zones, nature reserves, landscapes for recreation or simply as residential areas. There are different priorities for different natures depending on how they are legally, morally and materially mobilised.

To that end, there are three interrelated challenges that I particularly wish to address and that are of more general or principal concern when conceptually connecting landscape, responsibility and temporality. I set these challenges within what Head (2014, p. 89) has identified as the broader issue: 'how to approach a future that is open, inherently uncertain and subject to processes only partly under human control'. In identifying these challenges, I attempt to extend further notions of responsibility and temporality in landscape research.

The first challenge is a hands-on issue: who bears responsibility for sustainability configurations in different sites, spaces and times? Currently, horticultural plants are considered the greatest single source of alien species introductions to Norway. This places considerable responsibility on the nursery sector, and this responsibility is now formally stated in the regulation under the Norwegian Nature Diversity Act. It is, however, surprising that the Fornebu project is promoted among landscape architects as an innovative project which meets future environmental challenges. In fact, *Sustainable Fornebu* was 'one of the winners of the prestigious European and Regional Planning Awards in 2014' for its combination of 'sustainable energy solutions with a broadly based strategy for safeguarding and strengthening the biological diversity and landscape qualities of the area' (Qvenild and Setten Forthcoming). It is hardly surprising that there is no mention of alien species.

What, then, are the relations between knowledge creators, brokers and users – when particular ecological knowledges are commissioned, generated, selected and acted upon, especially in culturally and/or legally powerful ways? For example, ecological knowledge concerning the plants likely existed before the legal boundary was set, so who is responsible for making sure the best available current ecological knowledge is brought to bear? Is it responsible to wait for something to be legally prescribed? Whose authority is needed to make a particular course of (in)action irresponsible? To paraphrase Adam and Groves (2007, p. xiv), we need to investigate in more detail the 'uneven relation between acting, knowing and taking responsibility'.

In the current context, legal responsibility is generally placed on environmental authorities, and more recently on the greenery sector. Moral responsibility concerns all of us. Fundamental to both ways of understanding responsibility is that 'liability and blame depend on knowledge: first, knowledge of a timeline of events leading up to damage being done, and secondly, knowledge on the part of whoever played a causal role in the timeline' (Adam and Groves 2007, p. 143). The alien species debate is, however, characterised by fuzzy timelines as well as fuzzy spatialities, creating fuzzy knowledges, which at Fornebu were the root causes of the conflicting views (Qvenild 2014).

This leads to the second challenge: the nature of responsibility when we recognise that shifting temporal and spatial contexts can introduce unplanned forms of culpability. At Fornebu, it proved challenging, even impossible, for Statsbygg to know which plants were 'safe' to use in the project as the categorisation of 'aliens' is likely to cover increasing numbers of species in the future (Qvenild 2014). So, crucially, folded into current species politics is, in fact, an institutionalised irresponsibility. Still, landscape architects and planners carry a responsibility for shaping environmentally sustainable landscapes which are designed to protect what at present is seen as valuable biodiversity, while also catering for an unknown future biodiversity. How can anyone act responsibly when an ideal within the environmental management sector is to freeze nature – by fixing and purifying it – while the contexts for making sustainable decisions are changing – for example, when species are being listed and when the lists are revised? By implication, the nature of environmentally sustainable landscapes is a moving target and this challenge calls for some elaboration.

Massey has usefully pointed out that 'The stake is not change itself ... for change of some sort is inevitable; rather it is the character and the terms of that change. It is here that the politics needs to be engaged' (2006, p. 40). This is a call for more explicit reflections over temporality, the nature of the future and why the future matters when understanding and acting in the present.

According to Anderson (2010) and Westholm (2012), 'the future' is a preoccupation within environmental research and politics. Not least in sustainability debates and among scholars problematising environmental ethics, the future features prominently. According to Anderson, the future is not necessarily problematised as a category (see also Harvey 2015). The future is taken for granted yet at the same time seen as crucial for the questions we ask and the future environmental disorder we are expecting, for example, climate change and the increasing movement of species for biodiversity. The problem with the future is that there is no empirical evidence – the future is open and hence can *only* be open for *discussion*, not for drawing conclusions (e.g. Adam and Groves 2007; Head 2014). How, then, can we act when 'the problem' is not really here? This makes the future highly political – it 'is there for the taking, open to commodification, colonisation and control, available for exploitation, exploration and elimination' (Adam and Groves 2007, p. 13). In addition, and by implication, 'because the future cannot be known, responsibility tends to be pushed outside the frame of reference and concern' (Adam and Groves 2007, p. xiv. See Robin, this volume.).

The solution is to make the future present in effective ways, that is, to make a *certain* future present by naturalising it. For biodiversity, a key feature currently seems to be to nationalise species by categorising them into aliens or natives and by taking action accordingly. Such practices rest on images and ideologies of 'historical abundance and subsequent decline' (Alagona *et al.* 2012, p. 49) as well as of a vision of 'the future as entirely void (apart from the results of our actions)' (Adam and Groves 2007, p. 195). This forces us to address the critical question of the *when* of nature. So, as Castree crucially reminds us:

> If we have a propensity to spatialize what we call 'nature', we also temporalise it too. Indeed, each is the analogue of the other. If I ask the seemingly strange question 'when is nature?', it soon becomes clear that we think these days that it's ever more a thing of the past.

> (2014, p. 12)

Temporally, and if nature actually is a thing of the past, this signals linearity and non-pluralism rather than 'multiple rhythms, events and trajectories over different scales depicting a world composed of a multiplicity of forces and trajectories with the potential for differentiation' (Lorimer 2012, p. 596). It also signals, again according to Anderson (2010), a concern for how geographies are lived and made as futures are prophesied, imagined or regularised. So, folded into the naturalisation of the future is a process of constituting the landscape as equally *empty* of particular species and practices: 'Certain lives may have to be abandoned, damaged or destroyed in order to protect, save or care for life' (Anderson 2010, p. 780, see also Brown 2015; Head *et al.* 2015; see Adam and Groves 2007, for the notion of 'empty future'). Such anticipatory action calls for a fundamentally challenging ethical and 'responsibilising' question: 'How to act in a way that protects and enhances some form of valued life?' (Anderson 2010, p. 782).

By implication, then, we also need to pay attention to *how* the future becomes manifest in the present. Anderson offers important clues in this respect as he argues that we need:

> to attend to how futures appear and disappear; to describe how present futures are intensified, blurred, repressed, erased, circulated or dampened; and to understand how the experience of the future relates to the materiality of the medium through which it is made present, whether that be a graph or an affective atmosphere.

> (2010, p. 739)

Landscape is a key medium – for example, by establishing buffer zones in order to prevent valuable lives from becoming extinct; and for landscape research, work needs to be done on how exactly the landscape makes the workings of an open and fluid future visible and powerful.

Following from this, the third and final challenge is to respond to the question of how 'landscape' informs a debate about responsibility – on the ground as well as analytically. How can the fixing of responsibility relate to it? The response to these questions very much hinges on how landscape is conceptualised and, by implication, mobilised in various contexts and to what ends. I want to end by highlighting and querying three strands of landscape research which in different ways hamper yet also have the potential to make the workings of landscape, responsibility and temporality a powerful perspective on a 'multinatural' future.

The first strand of research, with which I started, conceptualises landscape as fundamentally political, material, spatial and temporal, where temporality has to a large extent been confined to the past and the current (e.g. Bender 2002; Harvey 2015). By implication there is a vast landscape literature where the logic of a past is mobilised in the present (e.g. Setten 2004). By further implication, past time is taken for granted. This is a political debate concerning responsibility and the sociality and materiality of (future) landscapes. Such a debate needs, however, to interrogate more explicitly and critically how the temporal consequences of various (dis)connections and mobilisations of representational and material landscapes lead to greater understanding of how (ir)responsibility is recreated and sustained (see Brown 2015).

A second strand of landscape research has developed along non-representational and phenomenological perspectives (Harvey 2015; see Wylie 2007). Temporality is a major concern for these perspectives yet does not easily relate to responsibility. To that end, Harvey (2015) makes two highly relevant points – first, 'an unintended presentism, which through its excitement over living-in-the-moment seems to ignore a more complex and nuanced temporal perspective' (Harvey 2015, p. 3). With phenomenology's emphasis on the 'now', what about context, contingency and temporal accountability? Second, and related to responsibility, Harvey states that:

> the impossibility of 'speaking the truth' when attempting to represent the world can lead to a situation where nothing of consequence can be said about anything, and where the self becomes the only element that can be safely talked about.
>
> (2015, p. 3)

This is a perspective where time – as presence – plays a major role in how landscape is mobilised. What can we learn about responsibility when time has limited duration and where the landscape is down to only me? Harvey (2015, p. 11) reminds us that 'there is temporal depth and connection, even in the so-called fleeting and momentary'. The issue at stake here is to tease out how spatial and temporal responsibility can relate to presentism, yet also point to the future.

The third and final strand of research draws fundamentally and almost exclusively on the visual qualities of landscape, tending to gloss over other things (Setten and Brown 2009). Landscape imaginaries, such as scenarios, feature prominently in applied policy research and planning. Future landscapes become imagined as if they were actual or real, that is, they are already taking place and they are 'made to move and mobilize' (Anderson 2010, p. 785). Scenarios as visualised and materialised futures become tremendously powerful tools when politically enacted because visualisation is such a prominent and hence effective tool in bringing a future (landscape) into being, coupled with the open, abstract and fluid nature of the future. One of the dangers of scenarios lies in their ability to condition and limit intervention in the future (Anderson 2010, p. 785). So, ironically, scenarios might work against open futures. This forces reflections on the role of responsibility and whether and how to mobilise responsibility *for* futures.

It is time to return to my grandparents' *Rosa rugosa*. Should I remove it, should I manage it, or should I leave it? All options, however different they might seem, speak to responsibility, landscape and temporality. This shrub helps to root me somehow – emotionally, spatially and materially – in a world that hesitates to stand still. A plant can reveal to us that Nature and Society are never worlds apart, yet responsibility for their relationship is only ever the work of humans.

Acknowledgements

This chapter is based on a keynote speech at the Nordic Geographers Meeting in Reykjavik in 2013. A number of people have contributed to the reflections conveyed, and I want particularly to thank Marte Qvenild, Katrina M. Brown, Tomas Germundsson, Tom Mels, Don Mitchell and Edda Waage for their input.

References

Adam, B. and Groves, C. 2007. *Future Matters: Action, Knowledge, Ethics.* Leiden, The Netherlands: Brill.

Alagona, P.S., Sandlos, J. and Wiersma, Y.F. 2012. Past imperfect: Using historical ecology and baseline data for conservation and restoration projects in North America. *Environmental Philosophy*, 9(1), 49–70.

Anderson, B. 2010. Preemption, precaution, preparedness: Anticipatory action and future geographies. *Progress in Human Geography*, 34(6), 777–798.

Atchison, J. and Head, L. 2013. Eradicating bodies in invasive species management. *Environment and Planning D: Society and Space*, 31, 951–968.

Bender, B. 2002. Time and landscape. *Current Anthropology*, 43(S4), 103–112.

Brown, K.M. 2015. The role of landscape in regulating (ir)responsible conduct: Moral geographies of the 'proper control' of dogs. *Landscape Research*, 40(1), 39–56.

Castree, N. 2014. *Making Sense of Nature.* Abingdon, England: Routledge.

Gederaas, L., Moen, T.L., Skjelseth, S. and Larsen, L.-K. 2012. *Alien Species in Norway – with the Norwegian Black List.* Trondheim, Norway: Norwegian Biodiversity Information Centre.

Germundsson, T. 2004. The landscape of Vittskövle estate – At the crossroads of feudalism and modernity, in *European Rural Landscapes: Persistence and Change in a Globalizing Environment*, edited by H. Palang, H. Sooväli, M. Antrop and G. Setten. Dordrecht, The Netherlands: Kluwer Academic Publisher, 245–267.

Harvey, D.C. 2015. Landscape and heritage: Trajectories and consequences. *Landscape Research*, 40(8), 911–924.

Head, L. 2012. Decentering 1788: Beyond biotic nativeness. *Geographical Research*, 50(2), 166–178.

Head, L. 2014. Living in a weedy future: Insights from the garden, in *Rethinking Invasion Ecologies from the Environmental Humanities*, edited by J. Frawley and I. McCalman. New York, NY: Routledge, 87–99.

Head, L., Atchison, J. and Phillips, C. 2015. The distinctive capacities of plants: Re-thinking difference via invasive species. *Transactions of the Institute of British Geographers*, 40(3), 399–413.

Head, L. and Muir, P. 2004. Nativeness, invasiveness, and nation in Australian plants. *Geographical Review*, 94(2), 199–217.

IUCN Council. 2000. *Guidelines for the Prevention of Biodiversity Loss Caused by Invasive Alien Species*. Gland, Switzerland: International Union for Conservation of Nature.

Jones, O. and Cloke, P. 2002. *Tree Cultures: The Place of Trees and Trees in Their Place*. Oxford, England: Berg Publishers.

Kålås, J.A., Viken, Å., Henriksen, S. and Skjelseth, S. (eds). 2010. *The 2010 Norwegian Red List for Species*. Trondheim, Norway: Norwegian Biodiversity Information Centre.

Lorimer, J. 2012. Multinatural geographies for the Anthropocene. *Progress in Human Geography*, 36(5), 593–612.

Low, T. 2001. From ecology to politics: The human side of alien invasions, in *The Great Reshuffling. Human Dimensions of Invasive Alien Species*, edited by J.A. McNeely. Gland, Switzerland: International Union for Conservation of Nature, 35–42.

Massey, D. 2004. Geographies of responsibility. *Geografiska Annaler*, 86(1), 5–18.

Massey, D. 2006. Landscape as provocation: Reflections on moving mountains. *Journal of Material Culture*, 11(1–2), 33–48.

Massey, D. 2011. Landscape/space/politics: An essay. *The Future of Landscape and the Moving Image*. Accessed June 30 2015. Available at https://thefutureoflandscape.wordpress.com/landscapespacepolitics-an-essay/.

Ministry of the Environment. 2007. *Strategy on Invasive Alien Species*. Accessed June 23 2015. Available at https://www.regjeringen.no/globalassets/upload/md/vedlegg/planer/t-1460_eng.pdf.

Mitchell, D. 2012. *They Saved the Crops: Labor, Landscape, and the Struggle Over Industrial Farming in Bracero-Era California*. Athens, GA: University of Georgia Press.

Norwegian Nature Diversity Act. 2009. Accessed June 30 2015. Available at https://www.regjeringen.no/en/dokumenter/nature-diversity-act/id570549/.

Qvenild, M. 2012. Native nature and alien invasions: Battling with concepts and plants at Fornebu, Norway, in *Eco-Global Crimes: Contemporary Problems and Future Challenges*, edited by R. Ellefsen, R. Sollund and G. Larsen. Farnham, England: Ashgate, 233–255.

Qvenild, M. 2014. Wanted and unwanted nature: Landscape development at Fornebu, Norway. *Journal of Environmental Policy & Planning*, 16(2), 183–200.

Qvenild, M. and Setten, G. Forthcoming. Locating value in the Anthropocene: Baselines and the contested nature of invasive plants, in *Locating Value: Theory, Application and Critique*, edited by G. Hoskins and S. Saville. Oxford, England: Routledge.

Qvenild, M., Setten, G. and Skår, M. 2014. Politicising plants: Dwelling and invasive alien species in domestic gardens in Norway. *Norsk Geografisk Tidsskrift-Norwegian Journal of Geography*, 68(1), 22–33.

Robbins, P. and Moore, S. 2013. Ecological anxiety disorder: Diagnosing the politics of the Anthropocene. *Cultural Geographies*, 20(1), 3–19.

Setten, G. 2004. The habitus, the rule and the moral landscape. *Cultural Geographies*, 11(4), 389–415.

Setten, G. and Brown, K.M. 2009. Social and cultural geography: Moral landscapes, in *International Encyclopedia of Human Geography*, edited by R. Kitchin and N. Thrift, Vol. 7. Oxford, England: Elsevier, 191–195.

Setten, G. and Brown, K.M. 2013. Landscape and social justice, in *The Routledge Companion to Landscape Studies*, edited by P. Howard, I. Thompson and E. Waterton. Abingdon, England: Routledge, 243–252.

Smith, D.M. 2000. *Moral Geographies: Ethics in a World of Difference*. Edinburgh, UK: Edinburgh University Press.

Smout, T.C. 2003. The alien species in 20th-century Britain: Constructing a new vermin. *Landscape Research*, 28(1), 11–20.

Staeheli, L. 2013. Whose responsibility is it? Obligation, citizenship and social welfare. *Antipode*, 45(3), 521–540.

Warren, C.R. 2007. Perspectives on the 'alien' versus 'native' species debate: A critique of concepts, language and practice. *Progress in Human Geography*, 31(4), 427–446.

Westholm, E. 2012. Miljöforskningens framtidsbilder, in *Att utforska framtiden*, edited by S. Alm, J. Palme and E. Westholm. Stockholm, Sweden: Institutet för Framtidsstudier, Dialogos Förlag, 91–109.

Widgren, M. 2012. Landscape research in a world of domesticated landscapes: The role of values, theory, and concepts. *Quaternary International*, 251, 117–124.

Wylie, J. 2007. *Landscape*. Abingdon, England: Routledge.

4   Presence of absence, absence of presence, and extinction narratives

Dolly Jørgensen

Am I truly the last?

The Last Unicorn – an animated fantasy film from 1982 directed by Jules Bass and Arthur Rankin Jr. based on the 1968 book by Peter S. Beagle – centres around what is identified as the last unicorn and the quest to find out if any other unicorns survive in the world. In the opening sequence, two hunters appear in the unicorn's forest and pronounce that they will not find game there because of the unicorn's protection, but that the unicorn had better stay put because it is the last. The unicorn then reflects on that information:

> That cannot be. Why would I be the last? What do men know?! Because they have seen no unicorns for a while does not mean that we have all vanished. We do not vanish! There has never been a time without unicorns. We live forever. We are as old as the sky, old as the moon. We can be hunted, trapped. We can even be killed if we leave our forests, but we do not vanish. Am I truly the last?
>
> (*The Last Unicorn* 1982)

Two important ideas appear in this soliloquy that carry through the story. First, there is the problem of seeing and knowing. 'Because they have seen no unicorns for a while does not mean that we have all vanished', the unicorn says. Later when a farmer and a cart driver, see her but think she is a mare because the horn is invisible to them, the unicorn ponders to herself: 'I had forgotten that men cannot see unicorns. If men no longer know what they are looking at, there may well be other unicorns in the world yet, unknown and glad of it'. The question is: Can one be sure that something is not there simply because one does not see it?

Second, there is the idea that a species lives forever. 'There has never been a time without unicorns', she says. The unicorn, as well as a harpy which the unicorn encounters at a circus, is immortal. The immortality of the unicorn is contrasted with the mortality of the human body when halfway through the story the unicorn is turned into a young woman in order to avoid detection. While there is an element of the fantastical in the immortality issue, there is also the more general idea of lack of change – that nature will remain as it always has been; in the unicorn's forest, it is always springtime and the hunters cannot kill the animals. In the unicorn's world, nature is constant and unchanging.

This chapter will address where these issues – the problem of not seeing at a certain time and the idea of a static nature over time – converge in two historical searches for the last: the European beaver (*Castor fiber*) in Sweden at the end of the nineteenth century and the thylacine (*Thylacinus cynocephalus*) of Tasmania in the twentieth. Significantly these searches appear on the fringes of the modern developed world: the northern forests of the northern nation of Sweden and the island of Tasmania off the southeastern coast of Australia. These mountainous, sparsely populated areas encouraged both physical and conceptual distance from city dwellers. They were remote and difficult to access. There were internal frontiers on these edges of Europe and Australia, long after the American frontier had been declared conquered (Griffiths 1997).

Standing at the edge of civilization is standing at the edge of the known. As such, people moving beyond the frontier's edge are always confronted with ignorance. Within the field of environmental history, managing the unknown has attracted attention as a way of explaining what people do in their environment and why (e.g. Uekötter and Lübken 2014). For example, Susan Herrington has explained how ignorance of Canadian forests was perpetuated by the convergence of practical challenges of getting precise numbers about the extent and type of tree cover with cultural ideas of 'a cornucopia of natural resources' hampering knowledge (Herrington 2014, p. 53). Until the forest could be conceptualised as a finite good, there was little impetus to quantify or conserve it. The same holds true with many marine resources, which, because of difficulties assessing fish stocks and exploring the marine environment, have historically often been thought of as limitless (Sparenberg 2014). Just as in the unicorn's forest where it is always spring and death never comes, unknown environments are often imagined as expansive and inexhaustible.

When places are difficult to access, the contents of those places – whether people, animals, plants or geographical features – can be difficult to know. Combating ignorance of place and its contents may require surveys, scouting parties, knowledge sharing and other data collection, but the knowledge gained is still only partial and inexact (Zilberstein 2013). Environmental data, in particular, is always based on collection at a certain place at a certain moment in time. Even large sample sets, which may be able to represent the most probable aspects of the environment in question, are not all inclusive. There are always outliers. Animal populations are particularly problematic to capture because animals are mobile, and surveys by necessity are time/space-bound. It is tempting to see the failure to record an animal in a survey as proof of its non-existence, but as the unicorn muses, 'Because they have seen no unicorns for a while does not mean that we have all vanished'. The presence of an absence does not necessarily equate to an absence of presence, but in some cases it does.

In the two histories presented here, I explore how the presence of an absence (no known animals) became understood over time as an absence of presence (extinction) through narrative. Swedish beavers and Tasmanian thylacines had both become rare and then finally unseen, which led some people to claim their extinction. Others, however, claimed that the animals had survived and that they continued to exist in the wild fringes beyond civilization. Contentious conclusions resulted from the uncertainty of knowledge and management of the unknown. Consensus on the extinction of the beaver was more easily reached than the thylacine, but in both cases, extinction narratives became fixed and paved the way for efforts to reverse the extinctions. These histories reveal how extinction narratives are built on the acceptance of presence of absence as a sign for absence of presence.

Beavers in the backwoods

The story of the European beaver is the story of a near tragedy. By around 1880, there were probably 1,200 beavers in all of Europe, and these were restricted to small pocket populations: one in Norway, one in the Rhine valley, one on the Rhône, one in Ukraine and a few in Russia (Nolet and Rosell 1998). Most people think there is one kind of beaver globally. Actually, the beaver of Eurasia (*Castor fiber*) and the beaver of North America (*Castor canadensis*) are different species. In fact, they are very different species. They have a different number of chromosomes (48 in the European and 40 in the North American). There is no known hybridization between the two species and it is assumed that they cannot produce viable offspring. They diverged genetically about 7.5 million years ago after the early beaver migrated from Asia to North America (Horn *et al.* 2011).

The decline in the population of European beaver in Sweden was a long time in the making. As early as 1756, the naturalist Nils Gissler, a student of Carl Linnaeus, expressed concern that the beaver was being overhunted in Sweden. Before 'they never caught all of the pairs in each place, and never touched the young', he wrote, but now 'they kill all they can' with the result that the numbers were diminishing (Gissler 1756, p. 221).[1] In the nineteenth century, authors continued to remark on the dwindling beaver numbers, although they always assumed that the beaver was still present on the outskirts of civilization. G. Swederus, author of a major book on Scandinavian hunting, noted that the beaver was found 'only in the northern part of the country and in the wild tracts' since the animal was being pushed there by growing colonization (1832, p. 133). Professor of Natural History S. Nilsson (1832, plate 10) wrote that the beaver was then found only in the northern half of the country and 'there is no place that he is numerous, and he seems to become more rare each year. ... A generation ago, one found them there [in Jämtland/Norrland] in smaller colonies of 12-16 individuals; now one finds never more than a pair together, or a female with her young'. In these remarks, there was a sense of environmental change, yet the beaver was assumed to still exist somewhere in Sweden. Just as there had never been a time without unicorns in *The Last Unicorn*, in Sweden, there had never been a time without beavers.

Increasing rarity might be easy to recognise, but what about admitting something is no longer present at all? In 1873, Ferdinand Unander, the head of the agricultural school in Västerbotten County, wrote an article for the *Svenska Jägarförbundets Nya Tidskrift* with the title (translated as) 'One of Swedish hunting's lost precious animals' about the beaver. He noted that:

> People outside of Norrland still believe that the beaver exists in the vast and forested regions of northern Sweden in single streams that have not yet been visited by man. This view finds support in the zoological handbooks, that this animal is found in our country in these wastelands, however nowadays these are in all directions traversed by both Sami and Swedes, sometimes even by some tourists, and are more known to the inhabitants themselves than you usually imagine.
>
> (p. 28)

Unander contrasted the idea that beaver must be in the unknown places with the fact that there were no unknown places anymore in Sweden. He questioned the standard assumption that the beaver must still exist on the fringes of civilization because he believed there was abundant knowledge of the animals on that fringe. Going through the evidence of beaver sightings, he found that the latest evidence was from the far north in 1864, although the journal's editor added a footnote that beaver was seen in Jämtland up to 1866. Unander concluded 'that as long as no proof is shown that beaver is found in the Swedish dominion and by which refute the before given facts and figures, he [beaver] must be regarded as an animal extinct from the Swedish fauna' (1873, p. 33).

Unander's article was the first definitive statement of the beaver's extinction from Sweden. It was based on a lack of evidence for beaver after 1864 as well as the belief that Sweden was not in fact an unknown wilderness. Unander made the claim that the countryside was known intimately by this time: because the countryside was greatly crossed by locals and visitors alike, their failure to report beavers meant that the beaver was no longer present. Knowledge of the rural spaces equated to knowledge of extinction. Those familiar with the north would continue to make this claim. Cultural historian Eric Modin, for example, wrote in 1911 that:

> The last wild living beaver, which was confirmed seen in Sweden, was observed in 1866 at Lake Juveln in Kall parish in Jämtland. Reports have been provided that the beaver survived even later in this landscape, where he in any case appears to have remained the longest, but these have not been confirmed. There is no hope that he, as was supposed, still survives in some unknown mountain region, so crossed and known as the Norrland wilderness and mountain regions are now.
>
> (p. 192)

Modin, like Unander before him, argued that the north was not unknown, thus the beaver's extinction could be documented. An absence of presence was confirmation of the presence of absence.

After Unander's 1873 article, there was a hunt to identify which animal had been the last beaver in Sweden. Stories upon stories were recounted about the extinction of beavers in different parts of the country. A long multi-part article published in 1884 recounted the beaver's extinction history in Sweden, including a detailed account of numbers of beavers hunted over time in each county and the last beaver sightings in each (Anonymous 1884). Careful documentation was made of the situation in which each beaver was seen and the authority who reported it. Although there were reports of beavers in the 1870s in Jämtland, the lack of a skin subjected to 'scientific verification' meant that 'a misidentification was possible' (Anonymous 1884, p. 142). Yet even without this scientific evidence, the report concluded that beavers had probably been in Jämtland even in the 1870s (Anonymous 1884, p. 143). According to the director of the Jämtland county museum Eric Festin (1922), Modin was right that the last beaver in Lake Juveln was taken in 1866, but the last beaver in Sweden was killed 'according to reports' by a local man, Abraham Abrahamsson, in 1871. Festin (1922, p. 58) noted with great irony that this was 2 years before the total ban on beaver hunting in Sweden in 1873! Some accounts hinted that beavers might have lived on much longer than 1873. Sven Ekman (1910, pp. 211–212) wrote a book about hunting and fishing in northern Sweden, which included several pages about the extinction of the beaver, including sightings in the 1880s. Festin (1921, p. 148) also mentioned later sightings, noting that some claimed a beaver died as late as 1892, although it was 'not certain'. In these reports there is a clear questioning of the evidence – how much confirmation should there be to prove the beaver was still alive? Or more pressing, can lack of evidence mean the beaver is really extinct? As in *The Last Unicorn*, just because no person has seen it does it mean that there are no more?

Although there was controversy about which beaver had been the last, no one doubted that they were extinct in Sweden. Yet the European beaver was not extinct everywhere – it lived on in the neighbouring nation Norway where it had been afforded protection earlier than in Sweden (Collett 1883, p. 44). Because beavers still existed somewhere else, they could be reintroduced from those populations (Figure 4.1). Along with his pronouncement that beavers were indeed extinct in Sweden, Unander proposed 'importation from abroad and reintroduction of beaver' (1873, p. 32). It would take 50 years for this proposal to become reality. By the 1920s, a concrete movement was afoot to bring the beaver back to Sweden (Festin 1921, 1922). The first Swedish release of beavers from Norway took place in 1922. Based on its success, reintroductions spread throughout central and northern Sweden with a total of about 80 animals released in 19 different locations (Fries 1940, p. 139–144).

In publications reporting on the reintroduction efforts, stories about the last beaver previously in Sweden were always highlighted in the discourse because the extinction provided the grounds for the action. So, for example, when Sven Arbman wrote about the first reintroduction, he framed it within an extinction story about the last:

Figure 4.1 A taxidermied beaver on display at the Västerbottens Museum, Umeå, Sweden, 2012. This beaver was caught locally and was one of the descendants of the beavers reintroduced from Norway.

There is beaver in Sweden, wild, free, Scandinavian beaver, since June 6th 1922, 3:30 in the morning. It is more than half a century since that could last be said. In 1871 the last was shot in a stream near Sjougdnäs.

(1922, p. 274)

Highlighting the previous last beaver gave these new beavers significance. It is particularly telling that Arbman, as well as all the others writing in the 1920s, chose the 1871 beaver as the last. Although more beaver reports trickled in from remote places in Sweden after this date, confirmation of sightings became an issue and the consensus was to discount these. The rejection of later sightings was rhetorical – it makes for a better story if the last beaver died before full Swedish protection in 1873.

Thylacines in the hills

Unlike the European beaver where the species became extinct only in some areas so that it could be reintroduced later from remnant populations, the thylacine of Australia (*Thylacinus cynocephalus*) died out completely. Or did it?

In a way similar to the Swedish beaver story, people had long noted the decline in thylacines, known colloquially as the Tasmanian wolf, tiger or hyena. In his *History of Tasmania* from 1852, John West predicated that 'as every available spot of land is now occupied, it is probable that in a very few years this animal, so highly interesting to the zoologist, will become extinct' (p. 323). In Richard Lydekker's book on Australian animals, he acknowledged that 'a relentless war of extermination' of thylacines by settlers protecting their sheep 'has resulted in the almost complete extinction of this, the largest of the Australasian Carnivores, in the more settled portions of the country' (1896, p. 152). The end of this sentence is worth noting – Lydekker assumed that the thylacine was only extinct from the 'more settled' part, implying that it could still be numerous in other areas.

When photos of a thylacine in the Beaumaris zoo in Hobart were published in January 1934, the zoo held the only captive specimen in the world (Fleay 1934). Thylacines had been held at 13 zoos globally, including the Smithsonian National Zoo in Washington DC and Zoological Society of London, but those specimens had all died. The Hobart thylacine died on 7 September 1936. In March 1937, the City Council of Hobart offered £40 to anyone who could bring in a live thylacine in good condition for the zoo (*Examiner* 1937). The zoo had to quickly retract the offer, however, as it did not have the required permits from the Tasmanian Animals and Birds' Protection Board (TABPB, later to become the National Parks Service) which had approved a complete ban on thylacine hunting in 1936, only a few months before the Beaumaris zoo specimen died. While newspaper articles admitted that the thylacine was 'rapidly becoming extinct' (travelling correspondent to *The Mercury* 1937), it was claimed to still exist in the mountains and could thus be captured to repopulate the zoo.

The TABPB undertook expeditions to find thylacines in light of its ban on hunting and the demands for new zoo specimens. An expedition to count the thylacines in the mountainous region was organised in 1938 and a report of that search was published in 1939 (Sharland 1939). The report included descriptions of the search and photographs showing the expedition making plaster-of-paris casts of footprints of the thylacine, as well as recording other evidence of thylacine presence. No thylacines themselves, however, were spotted. This did not deter Sharland from believing that the species still survived: 'It must be emphasised, however, that its failure to reveal itself more frequently is not necessarily indicative of approaching extinction. Great areas of this game country are devoid of human inhabitants, while others are only sparsely inhabited' (Sharland 1939, p. 20). Because of the inaccessibility of the terrain and the tendency for animals to see a man first and 'get out of his way without itself being seen', Sharland believed the thylacine was still present in the remote areas of Tasmania (1939, p. 20). The conclusion of the report was that thylacines were indeed still present in the wild and thus sanctuaries should be declared in areas with encroaching settlements to protect the remaining thylacines (Sharland 1939, pp. 36–38).

Newspaper accounts over the next 50 years sporadically reported thylacine sightings, footprints and other evidence from the remote countryside. There was still enough faith that the thylacine existed and that the Taronga Zoological Park in Sydney received a permit from the Tasmanian Animal and Bird Protection Board to catch a pair of thylacines for conservation breeding purposes in 1949 (*Mercury* 1949). The hunt for the thylacine in February 1949 on the west coast of Tasmania turned up empty-handed (*Examiner* 1949).

In July 1953, the Australian naturalist David Fleay, the same scientist who had provided the photographs of the last known living thylacine from the Hobart zoo, asked, 'Has the remarkable marsupial wolf finally become extinct?' (p. 7). He noted that he had been on an expedition in 1945–1946, which managed to track down 'but one solitary individual in four months' intensive search only to have the animal escape the night it was caught in a carefully laid trap (Fleay 1953, p. 7). Fleay's article was a sobering report, yet he still held out hope that 'a philanthropist really interested in our much-vaunted fauna' could save the few remaining animals (1953, p. 7).

Narratives of extinction would be countered by newspaper proclamations like 'Tasmania's Tiger is not extinct!' (*Argus* 1957) and 'Wolf may not be extinct' (*Canberra Times* 1968) throughout the second half of the twentieth century. A strong claim was that the searches for thylacines, which had concentrated on the west coast, were looking in the wrong place. Some claimed to have seen thylacine tracks and witnessed baits taken from snares along the northeast coast (*Examiner* 1953). Even at the end of the twentieth century, some people still argued that the thylacine was present in the remote northeastern region; there is even a website cataloguing sightings in the area from 1953 to 1990 (Emberg and Emberg 2001) (Figure 4.2). Officials likewise continued to hold out hope. The Sydney Zoo again applied in 1954 for a permit to catch a thylacine. The permit was denied, but not because the thylacine was extinct but because 'the capture of the rare animal' was inappropriate (*Mercury* 1954, p. 3). The World Wildlife Fund organised a new search in 1980 (Brass 1980), and the media magnate Ted

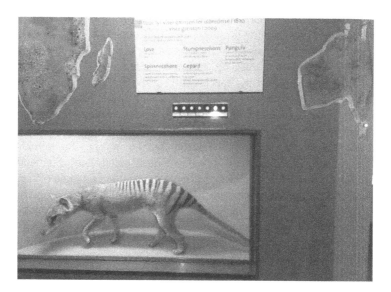

Figure 4.2 The thylacine at the Natural History Museum, Oslo, Norway, 2014. This display features a map with lights to indicate former and present (as of 2009) ranges of extinct and vanishing species. When the thylacine (*pungulv*) button is pushed, a green light appears in Tasmania, indicating that the thylacine is still present there, although by 2009 it had long been declared extinct by the International Union for the Conservation of Nature.

Turner offered $100,000 for a proven thylacine sighting in 1984 (*Canberra Times* 1984). Neither search found any animals.

The thylacine was eventually declared extinct by the International Union for the Conservation of Nature in 1982 and by the Tasmanian government in 1986 – the absence of presence was officially declared as a presence of absence. In the wake of this move, narratives to name the last Tasmanian tiger appeared. The narrative of extinction adopted in official documents has converged upon the identification of the Hobart zoo thylacine as the last. In 1996, Australia established National Threatened Species Day to commemorate the 60th anniversary of the death of the Hobart thylacine (Australian Department of the Environment and Heritage 2003). In many official statements about National Threatened Species Day, this zoo animal is called the 'last Tasmanian tiger' (Queensland Department of Environment and Heritage Protection 2014; New South Wales Office of Environment and Heritage 2015). The National Museum of Australia presents the thylacine as becoming extinct as a species on 7 September 1936 with the death of the Beaumaris zoo specimen, which has become known as 'Benjamin' although there are many questions about where this name originated (see Paddle 2000, pp. 198–200, for the controversy). When the Tangled Destinies exhibit, now known as the Old New Land exhibit, opened at the museum in 2001, Benjamin was described as the 'endling', the last of a species, in the exhibit text and pedagogical material accompanying it (Lewis and Arnold 2001). News stories about National Threatened Species Day also say the species became extinct with the death of Benjamin (e.g. Raabus 2007).

The narratives have converged on 7 September 1936 as the date of the thylacine's extinction, despite reports of animal sightings in the wildlands of Tasmania long after that date. Like the Swedish beaver extinction, this choice has narrative power because of two aspects to the story: first, the Tasmanian government had protected thylacines only 2 months before the individual's death; and second, the animal apparently died from cold exposure through neglect. Both aspects stress the avoidability of the extinction and make the extinction visible and deliberate.

Such a rhetorical position is useful for those who are working to bring the thylacine back from the dead. Not long after the extinction was declared, interest in resurrecting the thylacine through genetics appeared. The Evolutionary Biology Unit of the Australian Museum proposed in 1999 to extract DNA from a preserved thylacine pup in the museum archives. The grand idea was to reconstruct the thylacine genome and then implant a cloned embryo into a surrogate species. The team did manage to replicate individual gene fragments in 2002. But the project ended in 2005, primarily due to the poor quality of available DNA (see Fletcher 2008; Turner 2007). Yet, the thylacine has become a poster-child for 'deextinction' possibilities, which reached fever pitch in 2013 after a large TEDxDeExtinction event in Washington DC. Rarely does a publication about using genetic manipulation to recreate extinct species fail to mention the thylacine along with the passenger pigeon and woolly mammoth (e.g. Sherkow and Greenly 2013). The Tasmanian tiger has been classified as having 'un-dead status' (Smith 2012). This is all the more true in the twenty-first century as the thylacine's extinction is classified as potentially reversible.

While there has been official agreement on the end of the thylacine, sighting reports continue to this day. The unicorn's statement haunts the thylacine story: 'Because they have seen no unicorns for a while does not mean that we have all vanished'. The remoteness of the geography of Tasmania calls into question the known and the unknown. The question of how to read the lack of a thylacine body and whether that means extinction has been heavily contested in the thylacine history.

Extinction stories and the unwillingness to let go

Northern Sweden and Tasmania may be on opposite sides of the globe, but their positions on the edges of civilization created remarkable similarities in knowledge about nature. As these two histories reveal, many thought of these places as wild and unknown. Any failure to find a particular animal in these places was attributed to the observers simply being in the wrong place at the wrong time. The populations of beaver and thylacines, while potentially under threat, were thought inexhaustible in these remote and unknowable locations. At the same time, these woodlands on the frontier were colonised and criss-crossed more and more often by hunters and trappers. The geographies and animals within were not as unknown as some contended, thus the failure to find particular animal species was a sign of its extinction.

Confirmation of extinction is difficult because it assumes that if humans do not see the animal or signs of living animals, then the species must be extinct. This may prove to be a wrong conclusion. The presence of absence is not always equal to the absence of presence. Rediscovery of supposedly extinct species has become a regular occurrence. Examples range from the Lord Howe Island stick insect (*Dryococelus australis*) believed extinct since 1930 until a colony was found in 2001, to the flightless takahē bird (*Porphyrio hochstetteri*) of New Zealand thought extinct in 1898 and then rediscovered after a planned search effort in 1948, to the coelacanth (*Latimeria* spp.) which had only been found in the fossil record 65 million years ago until a live one was discovered in 1938 (Nelson n.d.). The unicorn had warned that failure to see unicorns should not be interpreted as their failure to exist. People want to believe that species live forever, and in the face of extinction, sometimes they are right that a species has survived. Examples like these, however, make it even harder to accept a real extinction event, when there is a complete absence of presence.

In both histories in this chapter, persons converged on one extinction story that became the standard narrative – the last Swedish beaver died in 1871 and the last thylacine died in 1936. Both convergences happened 50–60 years after the death of the particular animal identified as the last. This delay reveals the difficulty of identifying the last of a kind as sightings continue to be reported. Decisions were made to reject later evidence, partially because the evidence was deemed 'unconfirmable' but also because it failed to fit the extinction narrative. Both stories have politically charged conservation messages: the last Swedish beaver is said to have died 2 years before the species was legally protected, and the last thylacine is said to have died of neglect in a zoo only 2 months after the species was protected. On a narrative level, these stories were powerful messages about the failure of political and social systems to protect animal species. Kathryn Yusoff observed that 'the last of a species marks a juncture; it is a *being that stands for non-being*' (2012, p. 589, emphasis in original). By narratively confirming that the last individual of a species has died, the animal should then change status; it should become a non-being that is lost for all time.

Yet staring at potential non-beings, humans of the twentieth and twenty-first centuries have shown a remarkable unwillingness to let go. Some have been reluctant to let time run out on seemingly extinct species. Like the thylacine's extinction which has prompted physical searches for survivors, the alleged rediscovery of the ivory-billed woodpecker (*Campephilus principalis*) in 1999, 55 years after the previous confirmed sighting in 1944, prompted an intensive multi-year search and scientific publications about the woodpecker's potential survival (Heise 2010). Although the search ended without positive confirmation, the U.S. Fish and Wildlife Service released a *Recovery Plan for the Ivory-billed Woodpecker*, rather than recognise the species as extinct (USFWS 2010). In other cases, there are searches for survivors to put into conservation breeding programs and later reintroduce into former habitats, like the European bison (*Bison bonasus*) (Deinet *et al.* 2013). Deextinction efforts using genetic recovery and cloning or back-breeding are being used not only for the thylacine but also passenger pigeons (*Ectopistes migratorius*), gastric-brooding frogs (*Rheobatrachus* spp.) and others (Zimmer 2013). The common rhetoric of decline that accompanies species extinction in the modern era (Heise 2010) is countered by both rhetorical denials of extinction and concrete actions to bring species back.

The unwillingness to let a species go can have tangible effects on conservation actions. The amount of time and money available for conservation activities is limited and therefore activities must be prioritised (Mace *et al.* 2007). For this reason, some conservation biologists have accepted that the goal of zero extinction is unrealistic (Bottrill *et al.* 2008), although others have vehemently objected to that position (e.g. Parr *et al.* 2009). Searches for remnant populations of thylacines or ivory-billed woodpeckers, as well as reintroductions and complicated genetic species reconstruction efforts, consume time and resources that possibly could have been reallocated elsewhere if the species was assumed gone forever. The decision to recover a lost species comes with a cost. The decision to bring a species back is also predicated on recognizing the animal's extinction in the first place.

The histories of the Swedish beaver and Tasmanian thylacine reveal that extinction events become real to us through the stories we tell. It is in narrative that a species' presence or absence is determined. In these stories, storytellers search the record to identify the last in order to come to grips with the finality of extinction, even if it ends up not so final after all.

Note

1 All English translations of Swedish sources are by the author. This research was made possible through the project 'The Return of Native Nordic Fauna' funded by the Swedish research council Formas.

References

Anonymous. 1884. Bäfvern (Castor Fiber), Linné. *Svenska Jägarförbundets Nya Tidskrift*, 22, 83–84, 136–143, 236–245.

Arbman, S. 1922. När bäfvern återinfördes i Bjurälfven. *Svenska Jägareförbundets Tidskrift*, 60, 274–280.

Argus (Melbourne). 1957. Crosbie Morrison hails the good news of Tasmania's wildlife, 12 January, p.17. Accessed 16 November 2015. Available at http://nla.gov.au/nla.news-article71776004.

Australian Department of the Environment and Heritage. 2003. *Threatened Species Day Fact Sheet*. Accessed 16 November 2015. Available at http://www.environment.gov.au/resource/protecting-australias-threatened-species.

Beagle, P. 1968. *The Last Unicorn*. New York, NY: Viking.

Bottrill, M.C., Joseph, L.N., Carwardine, J., Bode, M., Cook, C., Game, E.T., *et al.* 2008. Is conservation triage just smart decision making? *Trends in Ecology and Evolution*, 23(2), 649–654.

Brass, K. 1980. The $55,000 search to find a Tasmanian tiger. *The Australian Women's Weekly*, pp.40–41. Accessed 24 September 2015. Available at http://nla.gov.au/nla.news-article47229295.

Canberra Times. 1968. Wolf may not be extinct. 26 April, p.3. Accessed 16 November 2015. Available at http://nla.gov.au/nla.news-article107048547.

Canberra Times. 1984. Offer for wolf sighting. 1 January, p.3. Accessed 16 November 2015. Available at http://nla.gov.au/nla.news-article116382208.

Collett, R. 1883. Om Bæveren (Castor fiber), og dens udbredelse i Norge fordum og nu. *Nyt Magazin for Naturvidenskaberne*, 28, 11–45.

Deinet, S., Ieronymidou, C., McRae, L., Burfield, I.J., Foppen, R.P., Collen, B., *et al.* 2013. *Wildlife Comeback in Europe: The Recovery of Selected Mammal and Bird Species*. London, England: Zoological Society of London.

Ekman, S. 1910. *Norrlands jakt och fiske*. Uppsala and Stockholm: Almquist and Wiksells.

Emberg, B. and Emberg, J.D. 2001. Thylacine sightings 1953–1990 in areas of north eastern Tasmania adjacent to the Panama forest. Accessed 16 November 2015. Available at http://www.tasmanian-tiger.com/thylafiles.html.

Examiner (Launceston). 1937. Wanted, a wolf! 9 March, p.6. Accessed 16 November 2015. Available at http://nla.gov.au/nla.news-article52127793.

Examiner (Launceston). 1949. Unsuccessful search. 7 March, p.3. Accessed 16 November 2015. Available at http://nla.gov.au/nla.news-article52667880.

Examiner (Launceston). 1953. They looked in the wrong place for Tas. tigers. 19 September, p.4. Accessed 16 November 2015. Available at http://nla.gov.au/nla.news-article61095067.

Festin, E. 1921. Bäverns återinplantering i Jämtland. *Sveriges Natur*, 12, 148.

Festin, E. 1922. Fridlysning av Bjurälvdalens karstlandskap och återinplantering av bävern: En samtidig lösning av två viktiga naturskuddsfrågor. *Sveriges Natur*, 13, 32–62.

Fleay, D. 1934. Hobart zoological collection. *The Australasian*. 20 January, p.43 plus photo essay in Pictoral Section p. iii. Accessed 16 November 2015. Available at http://nla.gov.au/nla.news-article141398749.

Fleay, D. 1953. Has the remarkable marsupial wolf finally become extinct? *The Courier-Mail* (Brisbane). 1 July, p.7. Accessed 16 November 2015. Available at http://nla.gov.au/nla.news-article51090120.

Fletcher, A.L. 2008. Bring 'em back alive: Taming the Tasmanian tiger cloning project. *Technology in Society*, 30, 194–201.

Fries, C. 1940. *Bäverland. En book om bävern och hans verk.* Stockholm, Sweden: Nordisk Rotogravyr.

Gissler, N. 1756. Rön och berättelse om Bäfverns natur, hushållning och fångande. *Kungl. Svenska vetenskapsakademiens handlingar*, 17, 207–221.

Griffiths, T. 1997. Ecology and empire: Towards an Australian history of the world, in *Ecology and Empire: Environmental History of Settler Societies*, edited by T. Griffiths and L. Robin. Edinburgh: Keele University Press, 1–16.

Heise, U. 2010. Lost dogs, last birds, and listed species: Cultures of extinction. *Configurations*, 18, 49–72.

Herrington, S. 2014. The forests of Canada: Seeing the forests for the trees, in *Managing the Unknown Essays on Environmental Ignorance*, edited by F. Uekötter and U. Lübken. New York, NY: Berghahn Books, 53–70.

Horn, S., Durka, W., Wolf, R., Ermala, A., Stubbe, A., Stubbe, M. *et al.* 2011. Mitochondrial genomes reveal slow rates of molecular evolution and the timing of speciation in beavers (*Castor*), one of the largest rodent species. *PLoS One*, 6(1), e14622. doi: 10.1371/journal.pone.0014622.

Lewis, R. and Arnold, D. 2001. *Tangled Destinies: Exploring Land and People in Australia Over Time Through the National Museum of Australia.* Canberra, Australia: National Museum of Australia.

Lydekker, R. 1896. *A Hand-book to the Marsupialia and Monotremata.* London, England: E. Lloyd.

Mace, G.M., Possingham, H.P. and Leader-Williams, N. 2007. Prioritizing choices in conservation, in *Key Topics in Conservation Biology*, edited by D. MacDonald and K. Service. Oxford, England: Blackwell, 17–34.

Mercury (Hobart). 1949. Attempt to preserve native tiger. 19 February, p.8. Accessed 16 November 2015. Available at http://nla.gov.au/nla.news-article26500580.

Mercury (Hobart). 1954. We'll save that tiger. 27 August, p.3. Accessed 16 November 2015. Avavilable at http://nla.gov.au/nla.news-article27222708.

Modin, E. 1911. Bör ej något göras för bäfverns återinförade i vårt land? *Svenska Jägareförbundets Tidskrift*, 40, 192–194.

Nelson, B. n.d. Lazarus species: 13 'extinct' animals found alive. *Mother Nature Network*. Accessed 16 November 2016. Available at http://www.mnn.com/earth-matters/animals/photos/lazarus-species-13-extinct-animals-found-alive/rediscovered.

New South Wales Office of Environment and Heritage. 2015. Threatened Species Day. Accessed 16 November 2015. Available at http://www.environment.nsw.gov.au/threatenedspecies/ThreatenedSpeciesDay.htm.

Nilsson, S. 1832. *Illuminerade Figurer Till Skandinaviens Fauna*, vol. 1. Lund, Sweden: UTI Academie-Boktryckeriet.

Nolet, B.A. and Rosell, F. 1998. Comeback of the beaver *Castor fiber*: An overview of old and new conservation problems. *Biological Conservation*, 83(2), 165–173.

Paddle, R. 2000. *The Last Tasmanian Tiger: The History and Extinction of the Thylacine.* Cambridge, England: Cambridge University Press.

Parr, M., Bennun, J.L., Boucher, T., Brooks, T., Chutas, C.A., Dinerstein, E., *et al.* 2009. Why we should aim for zero extinction. *Trends in Ecology and Evolution*, 24(4), 181.

Queensland Department of Environment and Heritage Protection. 2014. National Threatened Species Day. Accessed 16 November 2015. Available at https://www.ehp.qld.gov.au/wildlife/threatened-species-week/index.html.

Raabus, C. 2007. Remember Benjamin on National Threatened Species Day. *ABC Hobart*. 7 September. Accessed 16 November 2015. Available at http://www.abc.net.au/local/stories/2007/09/06/2026235.htm.

Sharland, M.S.R. 1939. In search of the thylacine: Society's interest in the preservation of a unique marsupial, in *Proceedings of the Royal Zoological Society of New South Wales for the Year 1938–9*. Sydney, Australia: Royal Zoological Society, 20–38.

Sherkow, J.S. and Greenly, H.T. 2013. What if extinction is not forever? *Science*, 340, 32.

Smith, N. 2012. The return of the living dead: Unsettlement and the Tasmanian tiger. *Journal of Australian Studies*, 36(3), 269–289.

Sparenberg, O. 2014. Perception and use of marine biological resources under national socialist autarky policy, in *Managing the Unknown: Essays on Environmental Ignorance*, edited by F. Uekötter and U. Lübken. New York, NY: Berghahn Books, 91–121.

Swederus, G. 1832. *Skandinaviens Jagt: Djurfånge Och Vildfavel. Jemte Jagtlexicon*. Stockholm, Sweden: P.A. Norstedt & Söner.

The Last Unicorn. 1982. [film] Directed by J. Bass and A. Rankin Jr. USA: Rankin/Bass Productions.

Travelling correspondent to *The Mercury*. 1937. The Tasmanian tiger: Early accounts of peculiar characteristics. *The Mercury*. 8 June, p.2.

Turner, S.S. 2007. Open-ended stories: Extinction narratives in genome time. *Literature and Medicine*, 26(1), 55–82.

Uekötter, F. and Lübken, U. (eds). 2014. *Managing the Unknown: Essays on Environmental Ignorance*. New York, NY: Berghahn Books.

Unander, F. 1873. Ett från svenska jagtbanan försvunnet dyrbart djur. *Svenska Jägarförbundets Nya Tidskrift*, 11, 28–33.

U.S. Fish and Wildlife Service (USFWS). 2010. *Recovery Plan for the Ivory-billed Woodpecker (Campephilus principalis)*. Atlanta, GA: USFWS.

West, J. 1852. *The History of Tasmania*, vol. 1. Tasmania, Australia: Henry Dowling.

Yusoff, K. 2012. Aesthetics of loss: Biodiversity, banal violence and biotic subjects. *Transactions of the Institute of British Geographers*, 37, 578–592.

Zilberstein, A. 2013. The natural history of early northeastern America: An inexact science, in *New Natures: Joining Environmental History with Science and Technology Studies*, edited by D. Jørgensen, F.A. Jørgensen and S.B. Pritchard. Pittsburgh, PA: University of Pittsburgh Press, 21–36.

Zimmer, C. 2013. Bringing them back to life. *National Geographic*. April. Accessed 16 November 2016. Available at http://ngm.nationalgeographic.com/2013/04/125-species-revival/zimmer-text.

5 The view from off-centre

Sweden and Australia in the imaginative discourse of the Anthropocene

Libby Robin

The idea that the Earth has entered a new geological epoch beyond the Holocene is one of the 'big ideas' of our time. It has aroused debate in environmental scholarship in the sciences, humanities and the policy sector. Artists, performers and museums are all working in different ways to explore the challenge of living in the Anthropocene – some through public activism, others through international governance. Humanities scholars have been prominent in taking the idea of the Anthropocene beyond its origins in the global change community of Earth system scientists. Conversations around the Anthropocene, they argue, offer many suggestive insights towards a 'planetary imaginary' (Clarke 2015, pp. 151–152). Thinking as a global citizen demands a scale that is both vast and also personal and locally responsible. The Anthropocene is an epoch of increasing inequity. The global changes that affect global warming, for example, have emerged from only some human behaviours, yet the effects are often felt first by those who have not benefitted from the advantages of a fossil fuel economy. Redistributing wealth and enabling the rise of alternative, transformative forms of energy are just some of the moral and technical questions raised by global change.

In this chapter, I consider how the citizens of the planet might imagine a more equitable future and also work to minimise future carbon pollution and other damaging, irreversible changes to global systems, both physical and cultural. When Nobel Prize winning chemist Paul Crutzen coined the concept of the Anthropocene at a meeting of the International Geosphere Biosphere Programme (IGBP) (Steffen 2013), the discourse focussed on changes in biophysical global systems. The description of the Great Acceleration of change, typically presented as a suite of 24 hockey-stick curves portraying changes since the 1950s, was the beginning of an Anthropocene discourse that explicitly included both biophysical and socio-cultural changes (Steffen *et al.* 2015), but it was still framed in IGBP-style curves. The idea of the Great Acceleration gave the Anthropocene a very particular history in the second half of the twentieth century and moved it towards a more accessible and popular realm. Not everyone was discouraged by the technical format. Bestselling novelist Dan Brown, for example, conjured with the magic of a version of the Great Acceleration curves in his 2013 novel *Inferno* (Steffen *et al.* 2015, p. 3).

Global change scientists have welcomed the emerging conversations with the social sciences and the arts. Crutzen has been an enthusiastic supporter of creative artists who explore the Anthropocene through performance, museums and public exhibitions, seeing these as 'optimistic' engagements engendering hope (Schwägerl 2014, p. 233). Museums also enable cross-disciplinary conversations between science and artistic endeavour beyond the academy. For example, the Anthropocene Project at the Haus der Kulturen der Welt (HKW) Berlin in 2014 hosted a meeting of the Stratigraphy Commission of the Geological Society of London alongside its exhibition and 'Anthropocene campus' events (Scherer 2014).

Even before the stratigraphers decide if we have officially entered the new epoch (which is possibly as soon as 2016; Zalasiewicz *et al.* 2015), the Anthropocene seems 'an idea whose time has come'. The Human Age, where the world is 'shaped by us' (Ackerman 2014), opens up particular concerns for humanists. Is the Anthropocene hubristic? Does it reinforce ideas of human domination of the planet? Does it signal the end of natural resource management, or 'control' of nature? At the opening of a new environmental humanities initiative The Seedbox Collaboratory (Åsberg and Hedrén 2015) at Linköping University, UCLA professor of literature Ursula Heise explored the rise of a new environmental humanities scholarship that focusses on the Anthropocene. Heise singled out the particular contributions of Sweden and Australia to this conversation (personal observation 17 September 2015). A rise in planetary scales of consciousness and the acceleration of 'global' have given smaller nations access to a new, even bigger, intellectual pond. Well practised as they are in watching politics beyond the domestic, Australians and Swedes, as well as Belgians, Dutch, Swiss and others, are well represented in the new community of global change science, where ideas of the Anthropocene have arisen.

New ideas of the global emerge when the discourse of the environmental humanities begins from off-centre, in extreme environments like Sweden and Australia. In this chapter, I open with a discussion of how a planetary vision demands negotiation with geological time, not just the short time frames of human economies, and then point to the value of *peripheral vision* in understanding the global scales of the Anthropocene. *Comparison* and transnational awareness are 'bread and butter' for smaller nations, but are essential to building a fair and just planetary awareness. In terms of the distinctive contributions of Sweden and Australia to the debates, I discuss the crucial role *extreme environments* have played in national identities in these countries and also their styles of *creativity*: they *internationalise* and *innovate* through science and technology. Science meets the humanities in distinctive ways: through the *ecological humanities* in Australia, and the *humanistic or green humanities* in Sweden. Sweden and Australia make edgy contributions to the discourse of the Anthropocene in the new scholarly area of environmental humanities, in the artistic and creative projects that flow into the broader public arena through museums, in exhibitions and in events that engage with global change.

Getting beyond the present and thinking with geological time

'Something is profoundly wrong with the way we live today', historian Tony Judt wrote in 2010. 'Much of what appears "natural" today dates from the 1980s: the obsession with wealth creation, the cult of privatization and the private sector, the growing disparities of rich and poor. And above all, the rhetoric which accompanies these' (pp. 1–2). Judt's concern was with social justice, rather than that of more-than-human nature, but he pinpoints the short-sightedness of a dominant way of living that dates back only 30 years. Time is compressed and the present is eternal. Profound long-term social and planetary changes are masked by this presentist mind set. The Western orthodoxy that focusses on growth at the cost of sustainability is not only entrenched in Western modes of living but is rapidly expanding to developing nations.

As a *human* epoch, the Anthropocene challenges include social equity, justice for all humans and for 'more-than-human others', even as we live in a world where biophysical systems have changed beyond any earlier point in the history of humanity on the planet. Moving beyond the Holocene puts us in a 'no analogue' world (Steffen *et al.* 2015, p. 14). While natural scientists writing about the Anthropocene tend to concentrate on stratigraphic debates about markers for a new geological epoch and questions of origin (Robin 2013, pp. 331–332), artists and creative writers have picked up on its value as a metaphor in emotionally stressful times of rapid change. Metaphor is an important way to engage with abstract ideas that function on larger-than-human scales, something Canadian geographer Brendon Larson has explored in the context of sustainability (Larson 2011). Climate scenarios being issued by the Intergovernmental Panel on Climate Change (the IPCC) and other reputable scientific groups generate widespread concern, but can also disempower individuals, businesses and governments (including small local governments) from engaging with the practical challenges of global warming. Alternative environmental narratives that tell global change in stories, in objects and in events can empower personal responses and enable cross-disciplinary thinking that is material, cultural and possibly game-changing.

Geographer Carl Sauer wrote about the 'meeting of natural and cultural history' in 1952, arguing that the global view of Earth is always personal, temporal and a view from *somewhere*. 'The things with which we are concerned are changing continuously and without end, and they take place, for good reason, not anywhere, but somewhere, that is in actual situations and places' (1952, p. 2). Sauer was interested in time in motion, 'changing in tempo', on many scales. His vision included a deep past and a long future. He was also attentive to the particularity of the present moment. He recognised its capacity to 'spread a veil' (Sauer 1952, p. 1) that disabled longer views.

Time is not equally deep everywhere. Time itself depends on place. Much of the dominant discourse of the era of postwar reconstruction, when both 'the environment' and the 'global' were conceived (Robin *et al.* 2013), was extrapolated from ideas of the national, particularly European and North American models.

This 'trans-Atlantic gaze' is based on assumptions from Western economies in temperate climates (Robin 2012). A default 'global' follows the big economies. It typically focusses on territories and land systems at the expense of atmospheres, oceans and polar ice-caps, which are the truly global places. The Australian continent has a distinctive deep-time history of cultures living through and surviving the past Ice Age. Massive sea level rises cut the land bridges to New Guinea and Tasmania and submerged much of the former continent ('Sahul'). While the 'civilization narrative' of Europe is that the ice melted and enabled the people to move north about 10,000 years ago, the Australian story is entirely different. There was very little ice, people created a civilisation that suited a low-energy environment and were living in the most arid parts of Australia right through the last glacial maximum, a cold and dry time (Robin 2012; Smith 2013). A new deep-time narrative has emerged in Australia from archaeological discoveries roughly in parallel with the new global change science that it now informs. Human history has had very different trajectories since people began to leave the 'cradle of humankind' in Africa. The undue focus on the epoch of the Holocene, when Western civilisation emerged and flourished, is a factor in the way we talk about the Anthropocene, as successor to the Holocene, yet it is not the only possible time scale to consider.

The musician Brian Eno took up the problem of planetary time in a different way. He understood that 'the precise moment you're in grows out of the past and is a seed for the future. The longer your sense of Now, the more past and future it includes' (2000). 'Now' is no longer just a moment. His work inspired the Clock of the Long Now, which 'ticks once a year, bongs once a century and the cuckoo comes out every millennium' a thinking device that embodies 'deep time for people' (Brand 1999, p. 4). Eno also challenged the Big Here, the global everywhere (that is going nowhere). The Anthropocene and the Long Now both came of age just as the new millennium began.

Transnational comparisons and the value of peripheral vision

In 'The spread of western science', a rare publication by a historian in the journal *Science*, George Basalla (1967) expounded a 'diffusionist' theory of scientific knowledge, which 'progressed' from the metropole to the periphery. Perhaps Basalla's theory with its nostalgia, parsimony and elegance held a special appeal for the editors of *Science*, sitting as they were at the centre of a knowledge web, at a time when scientific knowledge was expanding exponentially. Historians of science, particularly those working in 'peripheries' around the world, have been highly critical of the model (MacLeod 1980; Anderson 2002a). Despite the attraction of a simple narrative like that of Newton and his apple, it is often hard to find evidence that pinpoints where or when an idea began. New knowledge seldom rests on one datum point. The connectivities and recursive patterns of knowledge are such that the 'creative moment' is almost never static, singular or, in this well-connected age of Big Science, even the work of a single individual.

Already by the time Basalla was writing, the new knowledge revolution that followed the International Geophysical Year (IGY 1957–1958) was in full swing, very much at odds with his model. The IGY revealed the intellectual power of collective international effort and had a particular emphasis on far-flung places such as the Arctic, where Sweden was a leader of research, and the Antarctic, where Australia claimed 42 per cent of territory and was an important scientific power. The IGY inspired other intergovernmental science programs such as the UNESCO-sponsored Man and the Biosphere, established in 1971, which aimed to do for biology what the 'Big Science' approach had done for earth sciences. Coupled with the digital revolution, knowledge could no longer be characterised by linear models. International knowledge webs opened up different opportunities for nations on the edge.

'Small cultures' begin with the advantage that they already compare and know about other places, alongside their own. Swedish children watch subtitled Anglophone television before they can read, hearing English in their daily lives long before they learn it formally. Australian children of my generation played Monopoly on a board that gave them familiarity with the geography of London rather than any Australian city. Television has brought a familiarity with the favoured geographies of Hollywood to both Sweden and Australia. In these and many other subtle ways, we (of small cultures) have grown up unconsciously feeling that we live 'over there' rather than 'here', as we navigate parallel worlds of centre and periphery. The question of what small nations and middle powers can do in the world is important in both Sweden and Australia: there is a paradox that swimming as a lithe, mobile fish in a big pond may be more creative of national identity than growing fat in a smaller pond at home.

Sweden and Australia are of course not alone in 'speaking to the global'; the environmental humanities has flourished in other small rich nations too: a Swiss group, led by Marcus Hall, advocated the environmental humanities as a 'new window on grand societal challenges' (Hall *et al.* 2015, p. 134). A European Union group, led by Poul Holm, a Dane living in Ireland, argued for an 'Integrative Platform' for scholarship in the global humanities that brings together 'experts from all fields of science and scholarship to identify, review and develop current knowledge … and to identify what we know and what … we might know' (Holm *et al.* 2014, p. 191). Holm is a fisheries historian, and his global integrative platform was designed to serve particular questions of environmental and natural resource management. Sweden and Australia join other small nations, including Norway, in their concern to be relevant to the global. In Sweden and Australia, the environmental humanities have strongly overlapped with the emerging discourse of the Anthropocene, not least because science and technology studies have been prominent in both places.

National narratives from extreme places

Sweden and Australia share a national identity based on heroic relations with extreme places. Each has invested heavily in developing extreme environments – in Sweden, its Arctic and sub-Arctic north; and in Australia, deserts, the monsoon tropics and the frozen wastes of Antarctica. While desertification threatens global food security and the monsoon tropics are hotspots of biodiversity crucial to 'biosphere integrity' (Steffen *et al.* 2015), it is the polar environments that have become iconic in the discourse of global change science. The effects of global warming are magnified at the poles, while glacial ice cores have become what Australian historian Tom Griffiths has called 'the holy scripts, the sacred scrolls of our age' (2013, p. 359).

Sverker Sörlin has written eloquently of the narratives of the Arctic north that 'became prominent parts of Swedish self-understanding' (2002, p. 74) especially in the era following Norway's independence in 1905, when Sweden was establishing itself as a European industrial powerhouse. Sörlin uses the term 'internal colonialism' to describe the work of scientists who promoted economic expansion to increase population in the empty north and 'considered it as axiomatic that civilization must conquer all the world's wilderness' (2002, p. 90). It was 'irresistible' that an industrial higher culture would trump the nomads and farmers of the frigid Swedish north.

Australia's north was not cold but hot: tropical indeed. It still is the only developed Western nation with substantial lands in the tropics (38.6 per cent of its land mass or 2.97 million km², roughly six times the area of Sweden). As a 'frontier' for the white imagination, it was a place 'where civilization unravelled into wilderness – for good or ill' (Hains 2002, p. 5). These vast northern lands that hold just 1 per cent of the nation's population have an important place in the national psyche. As a frontier, the north has been a place of both anxiety and opportunity for a nation founded as a penal colony: it threatened regression but also offered renewal (Griffiths and Green 2011). The Australian north is littered with failed agricultural and pastoral schemes, and new ones are proposed again in the present day. Again and again, Australia has used the tools of science in efforts to 'civilize' its north, not through science for industry as in Sweden but through science for agriculture (Robin 2007). As Warwick Anderson has eloquently argued, the cultivation of whiteness in the north brought together science, health and destiny in Australia (2002b).

Both Sweden and Australia used their strong public science to grow their nations. Much of Australia's national scientific effort, particularly in the CSIR/O,[1] was put into agricultural expansion and 'improvements' in the deserts and the northern tropics. Settler Australians have 'battled' the land bravely, forging identity and entering global markets with wool, wheat and beef throughout much of the twentieth century. So entrenched has industrial agriculture become in Australia that, Lesley Head and colleagues argue, wheat is no longer considered a plant. It is merely a crop (Head *et al.* 2012). Cows are 'industrial' – so they do not belong in nature in Australia. By contrast, in Sweden, where modernity is framed by secondary industry, cows are an important part of the 'natural landscape' (Saltzman *et al.* 2011). Sörlin likened narratives of the Arctic North to Australian Indigenous songlines, which sing country into being, arguing that territorial claims and Swedish identity emerged out of the stories and images of 'natural historians, Arctic scientists, anthropologists' and other northern travellers (Sörlin 2002, p. 74). The comparison here is not between the imaginations of Indigenous peoples in each place but between the tropes of nationhood in two smaller nations anxious to use science to become leaders in a world bent on modernity.

Being global in the national interest

Sweden's long history of 'neutrality' and geopolitical independence reflects its need to have friends beyond superpowers. It is striking that the Swedish embassy was the third to be built (after the British High Commission and the US embassy) in Australia's national capital. It is a grand building showcasing the *avant garde* Swedish design and flair of 1947, typical of Swedish embassies all over the world, which marketed 'style' alongside diplomacy.

Travel for university and life education has a history dating back to the time of Linnaeus and his 'disciples' in Sweden. Australians too had to travel for doctoral degrees (usually to Britain, but sometimes the United States) until the Australian National University was established to award them in 1946, and other universities followed suit. The idea of being 'international citizens' is strong where travel is a normal part of education. Later in the twentieth century, the universities themselves became important in the economies of both countries through the export of university education. Swedes regard research and publishing in English as essential to being international, something attractive to international students. Nonetheless, publishing in English takes intellectual effort away from Swedish language and literature (Ekström and Sörlin 2012, p. 155) and remains a dilemma for Sweden and other small cultures with international aspirations.

Australians and Swedes also travel for work and host global projects. The role of the IGBP – based in Stockholm and for many years directed by Australian Will Steffen – has been very important in shaping the discourse of the Anthropocene. The *Future Earth* initiative, a new knowledge platform knitting together diverse global environmental change programs, has named five hub cities, including Stockholm (Future Earth 2015). Paul Shrivastava was named the director in February 2015. The chair of the Future Earth Science Committee is Australian CSIRO scientist Mark Stafford Smith, and his committee includes prominent Chinese-Australian expert on cities, Xuemei Bai, from the Australian National University (Future Earth Science Committee 2015). The Stockholm Resilience Center (SRC) founded in 2007 with the support of the Royal Swedish Academy of Sciences has been critical in shaping the language of global change science, particularly through integrative work across social, economic and ecological systems and concepts such as 'resilience' (Robin 2014) and 'planetary boundaries' (Rockström *et al.* 2009).

The nature of creativity and creativity with nature

Sweden and Australia both pride themselves on being 'creative nations'. Creativity is not just about artistic endeavour but also design and engineering (Sweden) and innovation (Australia). Science is an important creative force in both places, a public good, supported by the public purse. In Australia, where settler-farmers battled the land to grow food and fibre under strange ecological conditions, government science supported the project of 'improving' the country, and the emerging environmental problems of settlement became central to its focus (Robin and Griffiths 2004). Science also shapes the opportunities for humanities in both places, including expectations to be 'internationally' competitive and to have measurable outcomes.

In 1959, C. P. Snow declared that scientists have 'the future in their bones' (p. 6), and much of the authority to speak for the future throughout the twentieth century has been vested in science (Robin *et al.* 2013). But the discipline of Future Studies, something very strong in Sweden since the 1970s, was also closely associated with the humanism of the welfare state (Myrdal 1972). Myrdal's committee concerned itself with the social future, in an effort to bring 'acceptable material standards of living to all people'. It drew on science and technology, but tempered them with a concern not to create 'overhasty transformations and unforeseen negative effects' and recognition that the 'changes in social values and institutions' need to keep pace with technology (Myrdal in Robin *et al.* 2013, p. 273). There was a recent public review of the humanities in Sweden by Anders Ekström and Sverker Sörlin. Their report *Alltings Mått*, literally meaning 'everything measured', included a subtitle 'Humanistic knowledge for future society' (Ekström and Sörlin 2012). Humanist values are aligned with futures in Sweden, and 'science' already includes social and 'humanist' sciences.

In Australia, scientific institutions have also played a strong role in planning future visions, particularly the CSIRO and the Australian Science, Technology and Engineering Council (ASTEC, established in 1977). The Australian Research Council (ARC), established in 1988, following a major reform of higher education, remains a major public funder of scholarship. The ARC has framed a set of metrics directly from science. These often undervalue the style of the humanities, where books, for example, are more important than journal articles, but are not as easy to compare or collate into big data sets. The ARC metrics also value the international at the expense of the local and national. Any journal title including the word 'Australia' is rated lower than those with other national names (e.g. the British Journal of X, or the Journal of the American Academy of Y), even where the discipline in question is something like Australian Studies or Australian History.

There is a tradition, even perhaps an obligation in small societies like Sweden and Australia where university education is 'free',[2] to share the knowledge generated through research with the public. Public history, free legal aid and public interest policy studies have been very strong in Australia. In Sweden, academic staff are expected to give media interviews and contribute to society. There are not enough people in either place for specialists to become isolated from other disciplines, or hide away in an ivory tower. There is also a strong tradition of collaborative work across disciplines, for example, in the learned academies. Collaborative team research efforts with international partners are encouraged. Both Sweden and Australia have been leaders in 'cross-cutting' research in the 'post-disciplinary' mode; for example, Australians are well represented among students and research leaders at places like the SRC, and this Swedish initiative has been influential in Anthropocene debates. Public intellectuals are often included in media discussions and government advisory groups in times of crisis. Sweden even has a discipline of crisis studies, while Australia invests significantly in research into emergency management.

The ecological humanities and humanistic environmental studies

Conversations between physical sciences, particularly ecology, and the humanities have a long history in places where extreme conditions (e.g. fire and ice) shape life in obvious ways. Some key international scholars of environmental philosophy emerged in the 1970s and 1980s out of Australia (John Passmore, Peter Singer, Val Plumwood and Freya Mathews) and also Norway (Arne Naess). In 2001, the Australian National University hosted a new initiative in the ecological humanities, an interdisciplinary rapprochement between nature and culture, and between science and the humanities, at a time of 'environmental crisis' (Rose and Robin 2004; Griffiths 2015). Tom Griffiths identified three intellectual revolutions that drove this moment: the Darwinian revolution that made 'evolutionary history family history', new theories of chaos in ecology that eliminated the possibility of a 'landscape without humans' and Einstein's theory of relativity, where the mechanistic 'clockwork' universe of Newton was replaced by transforming energy from an 'elastic and viscous time-space' (Griffiths 2007). The ecological humanities initiative, which in early years included both Plumwood and Mathews, recognised Western, global imperatives but also considered the particular historical and ecological case of Australia, a settler society where the humanities encountered the 'bracing otherness of the natural world' (Griffiths 2015, p. 159). The ecological humanities initiative was built out of history, science and anthropology to meet an ethical imperative 'to be responsive to Indigenous people's knowledges and aspirations for justice … and to engage with connectivity and commitment in a time of crisis and concern' (Rose and Robin 2004).

Australia's ecological humanities thus began with the local and distinctive Australian ecological systems and focussed on connectivities between people and environments (including Indigenous perspectives, perceptions and emotional understanding of country through sense of place or belonging). Val Plumwood wrote of the 'ecological crisis of reason', not an environmental crisis, as she sought to include the more-than-human as holders of ethical rights (Plumwood 2002). Uniquely in the world, 'ecological' is also inclusive of public policy in Australia. Following the Rio conference in 1992, Australia developed a National Strategy for '*Ecologically* Sustainable Development' (ESD). The strategy explicitly included protection of biodiversity and ecosystem processes (as defined by ecologists, not economists), which has since been widely enshrined in Australian environmental law (Stein 2000). The ecological humanities embraced this language. Thus the future of nature is a physical and ethical question, not just about its value to the economy. ESD allows for the possibility of managing nature culturally (e.g. on Indigenous country) and reflected the 'Mabo decision', the important decision of the High Court (3 June 1992) that recognised native title for Aboriginal people and Torres Strait Islanders in common law. In many ways the ecological humanities addressed very specifically Australian policy making and philosophies and concerns from the 1990s era of Rio. The environmental humanities are more internationally inclusive and complement the planetary story of the Anthropocene that arose in the 2000s. When the new journal *Environmental Humanities* was launched in 2012 (Rose *et al.* 2012), the editors chose a title for an international readership.

The Environmental Humanities Laboratory is a project established in a history of science milieu at the Royal Institute of Technology (KTH), also in 2012, following the review of the Humanities in Sweden (Ekström and Sörlin 2012) and a generous donation by a Swedish businessman interested in promoting Swedish creativity. The idea of a 'Lab' picked up on the *experimental* nature of scholarship for a time when the rules of living with nature are no longer clear and when artistic, emotional, activist and scientific urges are in dialogue. The dynamism and energy of the Lab derived from an 'increasing demand on human and social knowledge to meet global challenges. The environmental humanities/*humanistiska miljöstudier* are characterised by a hybrid nature, combining skills, methods, and theories from several humanities disciplines ... and informed by science and technology' (KTH 2015). The Lab has been actively cultivating partnerships beyond Sweden, including sponsoring the journal *Environmental Humanities*. Sverker Sörlin, an architect of the Lab, is also an active researcher at the ecologically and globally influential SRC at Stockholm University and has appealed specifically to biologists to be more inclusive of environmental humanities knowledge (Sörlin 2012). Interdisciplinary collaboration, scientific team styles and creative endeavours have come together in 'the Lab' to reinvigorate responses to the crisis of accelerating material change and stagnating human responses to that crisis. Such conversations demanded a bilingual fluency in the languages of science and humanism. History of technology and 'science and technology studies' are strong in Sweden. This is also the disciplinary background of the curatorial team at the Deutsches Museum in Munich, Germany, where the first gallery of the Anthropocene opened in 2014 (Robin *et al.* 2014).

By contrast, in the North American context, the environmental humanities claimed a heritage in 'environmentalism' (social movements) and grounded its social imaginary in literature: the novel and literary non-fiction have been important sources for understanding global change, but what is meant by the 'environmental' has elided with the 'environmentalist' politic. Yet even among literary scholars science matters, as Ursula Heise has argued, focussing her work on the genre of 'science fiction' (2015).

The dance between science and humanities is different where outreach is mainstream work for scholars. In Sweden, where academic staff are expected to be public intellectuals, there is (as in Australia) less allegiance to an environmental movement (outside government) and more obligation to contribute directly to public policy and governmental debates. Only very recently have the Swedes argued that new questions 'can only be tackled by overcoming old boundaries' (Ekström and Sörlin 2012, p. 170). They have changed the parameters from futures 'management' to speak of handling crises through humanistic sciences (*humanistisk krishantering*) (Ekström and Sörlin 2012, p. 61). If 'the environment' is defined biophysically, the prediction of future environments is merely biophysical. But if, as Arjun Appadurai (2013) observed, the future is a 'cultural fact', then the human time frame is also important. The 'long crisis' of global change is hardly a *crisis* at all, Eric Paglia has argued (Paglia 2015). It demands a human-centred approach different from emergency management and 'triage' approaches (Soulé 1985), which cannot be maintained over the time scales of global change. Although these periods are but 'moments' in planetary history, they are experienced as long and slow in human time.

Edgy justice and cultural futures in the Anthropocene

Engaging the environmental humanities might enable ideas that can 'sing up' and reenchant our lands for an epoch where nature is no longer beyond our sense of self, borrowing an Indigenous way of speaking about land, as Deborah Rose (1996) explains. When you walk the songlines of Australia, you need to respect the ancestors, to approach all places with sacred reverence; the land is alive, and if you are an Aboriginal person, you belong to it (not the other way around). Aboriginal Australia's deeply civilised approach to country does not assume that nature is apart from humanity. Rather humanity is folded into nature. Good manners and respect for country require you to call out and establish your presence wherever you go. When you enter new places, you need to be welcomed. Nature has reciprocal moral rights, and country looks after humans where they care for it. In Europe and North American, wild nature and wilderness are precious, but increasingly managed by sleight of hand, protected aesthetically, if not actually, from human influence. If humanity is a global force, how can humans become planetary citizens? How might we learn to 'call out' imaginatively to the planet, enshrining an ethics of care for more-than-human others? How might individual humans care on a planetary scale?

Floods, fire and famine are all regarded as 'natural' disasters, yet they can be aggravated by human behaviours. Humanity still needs the physical world, and understanding how this works means engaging with the natural sciences as well as the ethical and humanistic ones. Environmental futures need both nature and culture. If accepted, the Anthropocene epoch will be marked by evidence in rock strata, likely to be traces of the nuclear explosions of the 1940s (Steffen *et al.* 2015; Zalasiewicz *et al.* 2015). It is not only about economics and the capitalist world system. The idea of 'Capitalocene', promoted by Alf Hornborg, Andreas Malm, James Moore and Donna Haraway (Haraway 2015, p. 163), brings important questions of economics and power to the Anthropocene, but robs it of its heritage in science, materiality and measurement. The 'Westphalian state' (market economies) is not the only thing at stake, as was recently suggested in *The History Manifesto* (Guldi and Armitage 2014, p. 76). The history of climate change is not just about Rachel Carson and the Club of Rome and other forces that motivated a politics of environmental change in the West; it is also about the changing climate and the history of the sciences based all over the world that learned to measure climate and weather through the World Meteorological Organization, the IGY and the IPCC (Robin *et al.* 2013). Only about 20 per cent of humanity is actually responsible for the emissions that cause global warming. Indeed, as Dipesh Chakrabarty argues, 'logically speaking the climate crisis is not inherently a result of economic inequalities' (2014, p. 11). The paradox is that it is *thanks to the poor* that the climate crisis is not worse. Planetary justice needs a basis in the physical world as well as the moral one. Climate and capital have 'conjoined histories' (Chakrabarty 2014).

Enlightenment thinking is planetary and human, not just Western, as Indigenous leader Noel Pearson has observed. There is a problem when the enlightenment becomes conflated with 'white fellas', when actually 'the enlightenment was a *human* achievement. It wasn't a western achievement or a British achievement or an English achievement' (Pearson 2014). Enlightenment traditions are important in the global imaginary, but they are not the sole property of the 20 per cent of humans whose emissions drive global change.

Beginning from a position 'off-centre' in framing the global scale raises different questions, even for nations like Sweden and Australia that contribute to these emissions. A well-known technique of artists seeking perspective on a grand landscape is to half close their eyes; without the clutter of too much information they can see and balance the essentials; they can look out for the big shapes in the scene they want to render. The view through the prism of a small nation's culture is oblique. An edge on the global allows a vision 'through a glass darkly'.[3] Because small nations are consciously comparative, they can construct perspectives in parallel and create a global imaginary that is informed by more than their own cultures.

The fact that 'almost every inch of the earth's surface is claimed and controlled by [Westphalian] states' (Guldi and Armitage 2014, p. 76) is actually beside the point, if the climate is changing and the very lands they claim are changing form and purpose. The global forces of political economy are important, but they are not the only game. The Anthropocene is not just about *causes*; it is about effects as well. The injustice of disadvantage created by global change is crucial. The paradox is that the epoch of the Anthropocene rarely offers solace, but it must still be lived in hope. This is where the arts and humanities can and must contribute.

Notes

1 The Council for Scientific and Industrial Research from 1926 to 1949; Commonwealth Scientific and Industrial Research Organization since 1949.
2 Most Australian university students do not pay upfront fees, but instead incur a publicly funded debt with low interest rates that becomes payable when income reaches a certain level.
3 The expression comes from 1 Corinthians 13:12, but was also famously used by Ingmar Bergman for the title of his 1961 film, in Swedish, *Såsom i en spegel.*

References

Ackerman, D. 2014. *The Human Age: The World Shaped by Us*. New York, NY: W.W. Norton.

Anderson, W. 2002a. Introduction: Postcolonial technoscience. *Social Studies of Science,* 32(5/6), 643–658.

Anderson, W. 2002b. *The Cultivation of Whiteness: Science Health and Racial Destiny in Australia.* Carlton, Australia: Melbourne University Press.

Appadurai, A. 2013. *The Future as a Cultural Fact: Essays in the Global Condition.* New York, NY: Verso.

Åsberg, C. and Hedrén, J. 2015. *The Seedbox Collaboratory: Research Program Overview and Current Project Portfolio.* Accessed 21 September 2015. Available at https://www.theseedbox.se/?l=en.

Basalla, G. 1967. The spread of western science. *Science,* 156(3775), 611–622.

Brand, S. 1999. *The Clock of the Long Now: Time and Responsibility.* London, England: Phoenix.

Chakrabarty, D. 2014. Climate and capital: On conjoined histories. *Critical Inquiry,* 41(1), 1–23.

Clarke, B. 2015. *Earth Life and System: Evolution and Ecology on a Gaian Planet.* Oxford, England: Oxford University Press.

Ekström, A. and Sörlin, S. (eds). 2012. *Alltings mått: humanistisk kunskap i framtidens samhälle.* Stockholm, Sweden: Norstedts.

Eno, B. 2000. The Long Now (filmed seminar). Accessed 21 September 2015. Available at http://longnow.org/seminars/02003/nov/14/the-long-now/.

Future Earth. 2015. Accessed 23 September 2015. Available at http://www.futureearth.org/ (see also ICSU, n.d. http://www.icsu.org/future-earth).

Future Earth Science Committee. 2015. Accessed 23 September 2015. Available at http://www.futureearth.org/science-committee.

Griffiths, T. 2007. The humanities and an environmentally sustainable Australia. *Australian Humanities Review,* 43. Accessed 21 September 2015. Available at http://www.australianhumanitiesreview.org/archive/Issue-December-2007/EcoHumanities/EcoGriffiths.html.

Griffiths, T. 2013. Commentary: Broecker (1987) and Petit *et al.* (1999), in *The Future of Nature: Documents of Global Change,* edited by L. Robin, S. Sörlin and P. Warde. New Haven, CT: Yale University Press, 359–362.

Griffiths, T. 2015. Environmental history, Australian style. *Australian Historical Studies,* 46(2), 157–173.

Griffiths, T. and Green, S.G. 2011. Culture, in *Australia and the Antarctic Treaty System: 50 Years of Influence,* edited by M. Haward and T. Griffiths. Sydney, Australia: UNSW Press, 346–372.

Guldi, J. and Armitage, D. 2014. *The History Manifesto.* New York, NY: Cambridge University Press.

Hains, B. 2002. *The Ice and the Inland: Mawson, Flynn and the Myth of the Frontier.* Carlton, Australia: Melbourne University Press.

Hall, M., Forêt, P., Kueffer, C., Pouliot, A. and Wiedmer, C. 2015. Seeing the environment through the humanities. *Gaia,* 42(2), 134–136.

Haraway, D. 2015. Anthropocene, Capitalocene, Plantationocene, Chthulucene: Making kin. *Environmental Humanities,* 6, 159–165.

Head, L., Atchison, J. and Gates, A. 2012. *Ingrained: A Human Bio-geography of Wheat.* Farnham, England: Ashgate.

72   *Libby Robin*

Heise, U.K. 2015. Environmental literature and the ambiguities of science. *Anglia: Journal of English Philology*, 133(1), 22–36.

Holm, P., Jarrick, A. and Scott, D. 2014. *Humanities World Report 2015.* London: Palgrave.

Judt, T. 2010. *Ill Fares the Land.* New York, NY: Penguin Press.

KTH. 2015. The Environmental Humanities Laboratory. Accessed 20 September 2015. Available at http://www.kth.se/en/abe/inst/philhist/historia/ehl.

Larson, B. 2011. *Metaphors for Environmental Sustainability: Redefining our Relationship with Nature.* New Haven, CT: Yale University Press.

MacLeod, R. 1980. On visiting the "moving metropolis": Reflections on the architecture of imperial science. *Historical Records of Australian Science*, 5, 1–16.

Myrdal, A. (Chair Swedish Institute). 1972. *To Choose a Future: A Basis for Discussions and Deliberations on Future Studies in Sweden* [*Att välja framtid*]. Stockholm: Swedish Institute.

Paglia, E. 2015. Not a proper crisis. *The Anthropocene Review*, 2(3), 247–261.

Pearson, N. 2014. Transcript *Q&A* 24 November (answer to question at 00:53:43). Accessed 21 September 2015. Available at http://www.abc.net.au/tv/qanda/txt/s4110610.htm.

Plumwood, V. 2002. *Environmental Culture: The Ecological Crisis of Reason.* London, England: Routledge.

Robin, L. 2007. *How a Continent Created a Nation.* Sydney, Australia: University of New South Wales Press.

Robin, L. 2012. Australia in global environmental history, in *A Companion to Global Environmental History*, edited by J.R. McNeill and E.S. Mauldin. Oxford, England: Wiley-Blackwell, 182–195.

Robin, L. 2013. Histories for changing times: Entering the Anthropocene? *Australian Historical Studies*, 44(3), 329–340.

Robin, L. 2014. Resilience in the Anthropocene: A biography, in *Rethinking Invasion Ecologies from the Environmental Humanities*, edited by J. Frawley and I. McCalman. London, England: Routledge, 45–63.

Robin, L. and Griffiths T. 2004. Environmental history in Australasia. *Environment and History*, 10(4), 439–474.

Robin, L., Avango, D., Keogh, L., Möllers, N., Scherer, B. and Trischler, H. 2014. Three galleries of the Anthropocene. *The Anthropocene Review*, 1(3), 207–224.

Robin, L., Sörlin, S. and Warde, P. (eds). 2013. *The Future of Nature: Documents of Global Change.* New Haven, CT: Yale University Press.

Rockström, J., Steffen, W., Noon, K., Persson, Å., Chapin, F.S., Lambin, E.F., *et al.* 2009. A safe operating space for humanity. *Nature*, 461, 472–475.

Rose, D. 1996. *Nourishing Terrains.* Canberra, Australia: Australian Heritage Commission.

Rose, D. and Robin, L. 2004. The ecological humanities in action: An invitation. *Australian Humanities Review*, 31–32. Accessed 23 September 2015. Available at http://www.australianhumanitiesreview.org/archive/Issue-April-2004/rose.html.

Rose, D.B., van Dooren, T., Chrulew, M., Cooke, S., Kearnes, M. and O'Gorman, E. 2012. Thinking through the environment, unsettling the humanities. *Environmental Humanities*, 1, 1–5.

Saltzman, K., Head, L. and Stenseke, M. 2011. Do cows belong in nature? The cultural basis of agriculture in Sweden and Australia. *Journal of Rural Studies*, 27, 54–62.

Sauer, C.O. 1952. *Agricultural Origins and Dispersals.* New York, NY: American Geographical Society.

Scherer, B. (ed.). 2014. *The Anthropocene Project: A Report.* Berlin, Germany: Haus der Kulturen der Welt (HKW).

Schwägerl, C. 2014. *The Anthropocene: The Human Era and How It Shapes Our Planet.* Santa Fe, NM: Synergetic Press.

Smith, M.A. 2013. *The Archaeology of Australia's Deserts.* New York, NY: Cambridge University Press.

Snow, C.P. 1959. *The Two Cultures.* London, England: Cambridge University Press.

Sörlin, S. 2002. Rituals and resources of natural history: The North and the Arctic in Swedish scientific nationalism, in *Narrating the Arctic: A Cultural History of Nordic Scientific Practices*, edited by M. Bravo and S. Sörlin. Cambridge, MA: Science History Publications, 73–122.

Sörlin, S. 2012. Environmental humanities: Why should biologists interested in the environment take the humanities seriously? *BioScience*, 62(9), 788–789.

Soulé, M.E. 1985. What is conservation biology? *Bioscience*, 35(11), 727–734.

Steffen, W. 2013. Commentary: Crutzen and Stoermer on the Anthropocene, in *The Future of Nature: Documents of Global Change*, edited by L. Robin, S. Sörlin and P. Warde. New Haven, CT: Yale University Press, 486–490.

Steffen, W., Broadgate, W., Deutsch, L., Gaffney, O. and Ludwig, C. 2015. The trajectory of the Anthropocene: The Great Acceleration. *The Anthropocene Review*, 2(1), 1–18.

Stein, P.L. 2000. Are decision-makers too cautious with the precautionary principle? *Environmental and Planning Law Journal*, 17(1), 3–23.

Zalasiewicz, J., Waters, C.N., Williams, M., Barnoskyc, A.D., Cearretad, A., Crutzen, P., *et al.* 2015. When did the Anthropocene begin? A mid-twentieth century boundary level is stratigraphically optimal. *Quaternary International*, 383, 196–203.

Part II
Living with nature in motion

6 The co-presence of past and future in the practice of environmental management

Implications for rural-amenity landscapes

Benjamin Cooke

Introduction

Environmental management research has become increasingly concerned with how interactions between rural-amenity migrants and the biophysical landscapes they come to occupy are shaping future ecological trajectories. Attention to these human–environment interactions is beginning to shed light on how environmental management 'practice' is not simply the deployment of human aspirations *in* the environment but a process of ongoing interplay between human and nonhuman agents (Gill *et al.* 2010; Cooke and Lane 2015a). I suggest that repositioning the on-ground practice of environmental management in a way that is attentive to nonhuman agency can encourage us to think differently about ecological change (Abrams *et al.* 2012) and open up an opportunity to think differently about time.

Through the tangible interaction that comes with environmental management practice, amenity migrants are perceptually engaged with the legacies of land use histories (Cooke and Lane 2015b). In Australia, rural regions that are experiencing amenity migration have been shaped by both pre- and postcolonial landscape modification of varying intensities. Through time, these historic interventions – like fires set by Indigenous peoples to manage the landscape or more recent clearing of forested lands for agricultural uses – have served to structure the ecological assemblages that now persist. Thus, when rural-amenity landholders engage in management practices like tree planting or weed removal on their properties, with the intention of creating new ecological assemblages, they are engaging with a legacy that remains active in the agency of plants, animals and soil, among other landscape aspects (Lien and Davison 2010). The way in which the past is brought into the present through efforts to make the future has implications for how we think about ecological transformation in rural-amenity landscapes.

Now is an opportune moment for critically examining the temporalities of environmental management, given the growing traction of the Anthropocene epoch in geography and ecology (Castree 2014; Ellis 2015). The Anthropocene disrupts precolonial and prehuman ecological benchmarks as the objectives for environmental management, emphasising that the 'thoroughly humanized earth'

(Castree 2014, p. 437) we now occupy will require new forms of open and reflexive environmental management that can accommodate future uncertainty (Ellis *et al.* 2012; Mastnak *et al.* 2014). I posit that efforts to progress a reflexive conception of environmental management in rural-amenity landscapes will need to engage with how embodied landscape histories structure future ecological trajectories and how the lived experience of ecological change over time will shape the receptiveness of human inhabitants to environmental management that consigns 'nature' to the Holocene. In this chapter I draw some insights from qualitative research with rural-amenity migrants in the hinterland regions of Melbourne, Australia, to contribute to contemporary discussions about how we might progress reflexive environmental management in the Anthropocene.

The socio-ecological implications of amenity migration

'Rural-amenity migration', that is, the migration of predominantly urban dwellers to rural landscapes of high 'amenity' value, is not a recent phenomenon. In Europe, North America and Australia in particular, the recreational, aesthetic and ecological qualities of rural countryside and coastal hinterland regions have been a draw card for well over half a century for retirees, second homebuyers and those seeking a lifestyle change (Curry *et al.* 2001; Van Auken 2010; Gosnell 2011). However, shifting patterns of global agricultural investment to developing nations, improved transport infrastructure to rural regions on the urban periphery and a growing appetite for rural residential development have seen an acceleration of existing trends. We are now confronted with a mosaic of different land uses and management intentions as consumption and conservation values are deployed alongside primary production, both across the landscape and within individual property parcels (Abrams *et al.* 2012). Accompanying this structural and demographic shift has been public and academic interest in the land use and ecological implications for rural landscapes – what is to be the future of the rural and hinterland regions that were once valued primarily for their production values?

At present, the dominant narrative around ecological implications for rural-amenity landscapes in Australia is one of threatening processes and ecological simplification (Argent *et al.* 2010; Cooke and Lane 2015b). Negative ecological impacts have been framed in terms of habitat fragmentation through the subdivision and settlement of rural and seminatural land. This fragmentation is occurring as larger properties are subdivided into smaller ones and more flora and fauna are lost to clearance for dwellings and the associated space needed for lifestyle activities (lawns, outbuildings). Negative implications have also been attached to amenity migrants themselves, with suggestions that as 'consumers' of amenity they are more likely to be interested in nurturing landscape aesthetics rather than managing for ecological function (Kendra and Hull 2005).

Opposing the narrative of negative impacts is the suggestion that positive ecological outcomes through rural-amenity migration can be achieved. In terms of physical landscape change, there is some evidence from landscape-scale analyses that vegetation cover has increased in rural-amenity regions as land uses become more diversified away from farming (Buxton *et al.* 2006). Through a combination of active restoration, regeneration and cessation of intensive primary production, forests have been slowly returning to grazing lands. Moreover, rural-amenity migrants themselves have also been noted to purchase their properties with the express intention of 'bringing back' nature to landscape that had been historically cleared for farming (Argent *et al.* 2010).

While the current narrative around the ecological implications of rural-amenity migration demonstrates the complexity of rural landscape change, research and practice must now move beyond the 'impact' mentality of human actors *on* the landscape, to a footing that locates people *in* the historical and biophysical contexts they inhabit (Trigger *et al.* 2008; Abrams *et al.* 2012). A more nuanced conception of these ecological implications is now needed to open up a wider discussion concerning how we address these challenges through environmental management practice and policy.

Nonhuman agency, temporality and environmental management practice

Exploring the intimate, fine-grained processes of environmental management practice on rural-amenity properties offers an avenue for unpacking how landscape temporalities interact with human agency to shape future ecological trajectories. Scholars have only recently begun to turn their attention to exploring how the 'complex dynamics of rural occupancy' (Holmes 2006, p. 156) are shaping rural-amenity landscape change, especially outside of landscape studies emanating from the United Kingdom and United States (Gill *et al.* 2010). As noted by Abrams *et al.* (2012), the emerging research on management practices presents an opportunity to probe the ways in which human and nonhuman agency are entwined in affecting ecological change. As has been usefully demonstrated in human geography, anthropology and archaeology in particular, the physical landscape is more than a blank canvas in which human activity is conducted, but an actor in its own right (Tilley 2004; Lien and Davison 2010; Ingold 2011). Due to the centrality of plants in the processes of environmental management practice and ecological restoration, I focus here on the agency of people–plant relations (Head *et al.* 2014). Plants in the form of the ecological assemblages that persist in rural-amenity landscapes are prominent embodiments of ecological histories, offering a way to explore how these histories influence current management practice.

The idea of a 'dialogue' between people and plants *over time*, where both entities are affected through the relationship, has proved perceptive in research into backyard gardens (Power 2005); the ongoing dimension of this dialogue is of particular relevance for the themes of this chapter. The work of Head and Muir (2006),

Power (2005) and Doody *et al.* (2014), among others, has shown how the inter-action between people and plants is an active 'conversation' that affects both gardens and gardeners (see Chapter 8). As people garden, the landscape form and arrangement of plants mediates the aspirations that residents bring to their gardens, connecting the history of past landscape modification to contempo-rary efforts to tend the garden. To reflect this conversational relationship for the remainder of this chapter, I define environmental management practice as 'any form of interaction between landholder and landscape – motivated by a conser-vation aspiration or broader amenity land use aspiration – that shapes ecological assemblages' (Cooke and Lane 2015b, p. 234).

The history of rural landscape modification on rural-amenity properties shares parallels with the backyard garden; people–plant relations of the past will shape a trajectory for the form and function of ecologies into the future. However, these pasts are not 'transmitted ready-made' (Ingold 2011, p. 141) into the present and future, but interpreted and enacted through landholders' lived experience and understanding of the ecologies in which they dwell. Indeed, perceptions of ecological function and appropriate environmental man-agement are indivisible from the time spent actively observing and interacting with those ecologies (Waage 2010). In the process of repositioning environ-mental management for the Anthropocene, it is vital that we interrogate how ecological futures are already being interpreted and co-produced at a fine-grained scale. An examination of environmental management that is conscious of temporal human–environment interactions may reveal useful insights into the challenges for environmental management practice and policy in rural-amenity landscapes.

Methodology

Study area

The site of this research project was the hinterlands of Melbourne, Australia. Melbourne's hinterland was chosen for this research due to the rapid pace of rural-amenity land use transition in this region over the last few decades (Mendham and Curtis 2010). Areas on the coast or coastal hinterland within commutable distance of Melbourne have proved most popular (Argent *et al.* 2010). My research focuses on two localities within Melbourne's hinterland – the Corangamite catchment (60 km west of Melbourne) and the Bass Valley district (80 km southeast of Melbourne). Recent research indicates that up to 50 per cent of the properties in parts of the Corangamite catchment are likely to change hands by 2020 (Mendham and Curtis 2010). The Bass Valley region experienced a 25 per cent increase in population between 1991 and 2006, making it one of the fastest growing regions on the fringes of Melbourne. The increasing rate of property turnover and changing land use away from intensive farming in these areas are symbolic of rural land use change globally (Gosnell 2011).

Research design and participants

A need to engage with participants in the environments where management was conducted guided this case study. To explore management practice, narrative interviews and a form of participant observation called the 'walkabout method' (Strang 2010) were conducted. The narrative interviews adapted an oral history style that encourages participants to tell stories about their experiences interacting with and observing their surrounding landscape over time. Participants were prompted for stories about their experiences living on their property, how the landscape had changed over time, and how their land management aspirations and practices had progressed. These interviews were conducted in or around the home of the participant and aimed to understand the environmental management aspirations they had when they arrived (if any), their early management interventions and what ecological changes those generated.

Following the interview, I walked participants' properties with them to explore how management practices had materialised in the landscape over time. The walkabout method was vital as it explicitly acknowledges that the physical environments that are of importance to people's lives will serve as repositories of memory, helping to trigger reflections on activities conducted in those spaces (Strang 2010; Trigger *et al.* 2010). As I walked the property with participants, evidence of land management embodied in the landscape (like tree planting, failed weed suppression) served as catalysts for stories about how those management practices were informed and conducted, how ecologies changed or responded to intervention, how subsequent practices might have evolved as a result and what had been learnt about ecological function through inhabiting the landscape over time. During the walks, photos were taken of management activities and ecological features encountered. The participant led the walk and told stories about management activities, though it was notable that plants played a powerful role in dictating our path – many landholders gravitated towards the site of rapid plant growth or ecological changes and five landholders veered from tracks around their property to remove weeds by hand as we walked.

Twenty-one landholders were interviewed between June and October 2010; details of these participants are shown in Table 6.1. The majority of landholders moved from suburban Melbourne, with three moving from smaller residential properties in rural townships and three had prior farming experience. The environmental management objectives of participants are noted in Table 6.1 to demonstrate the aspirations that they brought with them to their rural properties and serve as a reference point for the discussion of the management practices that later eventuated. Around half of the participants were recruited through their participation in local conservation programs, which suggests the cohort interviewed are more likely to be interested in conservation than landholders who do not engage with these schemes

Table 6.1 Profiles of participants involved in this case study and their management aspirations (names appearing here are pseudonyms).

Participants	Age	Formative environmental management aspirations that amenity migrants had for their property	Time on property
Beatrice and Jim	40–49	Conserve the existing forest on the property	15 years
Steve	50–59	Leave the forest to regenerate and planting understorey	9 years
Kelly	60–69	Conserve existing forest and plant out open areas around the house	11 years
Liz	50–59	Conserve existing forest and allow natural regeneration of paddocks to continue	20+ years
Rob	50–59	Leave forested area largely untouched	20+ years
Trevor	70–79	Leave paddocks intact and plant linear tree buffers to shelter livestock along the property boundary	20+ years
Alex and Simone	30–39	Leave forested areas intact and plant linear tree buffers around paddock fence lines	14 years
Emma	70–79	Establish garden and allow natural regeneration of former pine plantation	26 years
Sally	40–49	Leave forested area and establish a garden and lawn around the house	8 years
Karen	70–79	Replant understorey vegetation among scattered trees	13 years
Ken	50–59	Restore woodland to cleared, former grazing land	6 years
Maddy	50–59	Allow grazing land to regenerate and establish a garden	8 years
Alice and Sam	50–59	Remove weeds from creek line and plant trees in small patches	14 years
Pauline and Allan	40–49	Plant linear tree buffers to shelter livestock along the fence line	12 years
Dan	70–79	Plant linear tree buffers to shelter livestock and forest patches to increase overall vegetation cover	28 years
William	40–49	Retain existing forest and plant linear tree buffer along fence line	17 years
Lauren	40–49	Plant understorey in cleared open paddock	22 years
Hannah	40–49	Allow woodland to regenerate and establish garden around the house	18 years
Nick	50–59	Replant woodland that had been cleared, plant garden and orchard close to house	7 years
Jeff and Claire	50–59	Retain forest area and establish garden	14 years
Tina	50–59	Replant forest to former grazing land, establish garden and orchard	22 years

Adapted from Cooke and Lane (2015b).

Experiencing temporality in environmental management practice

Resurrecting the past – amenity landholder aspirations

Rural-amenity migrants brought a diversity of aspirations for environmental management on their properties. Tellingly, however, eight participants spoke of ecological restoration as a strong motive for inmigration. References to past ecologies through phrases like 'bringing back' (Jim, Martina) nature, or restoring 'what belongs' (Emma, Steve) showed how past benchmarks associated with a precolonial ecology were consciously drawn on as part of management objectives for the property. While these redemptive motivations were often held in conjunction with a desire to establish ornamental gardens and pursue nonconservation land uses, it was early attempts at precolonial ecological restoration that revealed how embodied landscape histories could prove a disruptive force for achieving management goals. In this section I draw heavily from three participant narratives in particular, to bring nuance to the framing of environmental management as a dialogue between people and plants; the implications of the practice of environmental management are explored later in the chapter.

Encountering landscape legacy through vegetation removal

Lauren's property in the Bass Valley district was characterised by open paddocks that were home to a small herd of cattle, with a fenced patch of bushland of around 5 hectares at the bottom of a steep hill. The section of bush closest to the paddocks had been partially cleared by the previous landholder; as Lauren understood it, the purpose of this clearing was to provide sheltering space for livestock during storms, explaining why much of the understorey vegetation was removed, while most canopy trees were retained (Figure 6.1). Without active grazing, Bracken fern (*Pteridium esculentum*) had established quickly in the niche left by the clearing, creating a homogenous understorey. Given its status as a pioneer species, this rapid establishment is a typical phase of bushland regeneration where ecological disturbance occurs.

While bracken is a widespread precolonial benchmark species in Australia, it has had (and continues to have) the status of 'agricultural weed' in many jurisdictions, given its habit of overtaking grazing space in paddocks, while also being potentially poisonous to livestock (DEPI 2007). As such, the plant's behaviour and traits rather than its origin have attracted the categorisation of 'weed', allowing for it to be removed by farmers without attracting punitive action. This status of agricultural weed, combined with its pioneering habit, means it is often maligned, even among people with strong conservation aspirations.

As the open patch in Figure 6.1 was adjacent to more 'intact' forest, Lauren wanted to plant out the area with native shrub species, to 'bring it up to standard'. As one of the few landholders who grew up on a farm, Lauren viewed bracken as a weed, with its dominance through this block considered to be an 'invasion' that was inhibiting revegetation opportunities. Even though Lauren had learned from her local nursery that bracken was native prior to pursing its removal, her strong sense that its presence indicated land that was poorly managed (perhaps due to her upbringing on a farm) saw her push ahead regardless. As such, a few years prior to my visit, Lauren and her husband began to clear patches of bracken, with the intention of sourcing a range of native shrubs from a local nursery for planting in the spring. However, shortly after clearing the bracken, Lauren observed that blackberries had begun to spread in from her neighbour's

Figure 6.1 Bracken fern that had been left by Lauren is visible centre image; after observing how blackberries moved in to occupy the space where bracken was removed, these patches were left as they were. A thicket of blackberry can be seen bottom right.

property, establishing in the space where bracken was now absent. As a declared noxious weed in Victoria, Lauren considered this outcome to be much less desirable than the presence of bracken, resulting in a decision to leave the remaining bracken in place to reduce further blackberry spread. As one of Lauren's early management interventions, this event had a powerful effect on her approach to management, as well as dictating a particular trajectory for management practice across the property.

As Lauren reflected on the unfolding management interactions that had occurred over time, she noted the landscape as an actor, suggesting that you 'can't work against the bush' in your management efforts. The history of landscape modification on this property – and surrounding properties – had meant that the intention to restore a precolonial ecology had actually resulted in an ecological assemblage that had strayed even further from that intended benchmark. Aside from Lauren, nine other landholders provided a conscious reflection on the idea that you cannot just turn back the clock when it comes to ecological restoration. Through tacit engagement with plants, many participants had come to realise that the *response* of the landscape to their active intervention was often unpredictable (Head and Muir 2006). Indeed, 11 participants had adopted a philosophy of 'do a little bit and see what happens' (Claire) as they grew increasingly conscious of the potential for their well-intentioned management interventions to generate undesirable ecological outcomes (Howitt and Suchet-Pearson 2006).

Experiencing ecological change – setting trajectories for future ecologies

As with most participants listed in Table 6.1, Beatrice and Jim had lived on their property through a large proportion of the millennium drought in southeastern Australia, which persisted from 1998 to the late 2000s. Through this period, many of the plants in their bushland suffered some dieback in response to the dry conditions. However, the shrub species Swamp Paperbark (*Melaleuca ericifolia*) suffered a particularly acute decline, given its preference for 'swampy' environments (Figure 6.2). Swamp Paperbark is a species that is considered to belong here as part of the local benchmark ecological vegetation community, but is sensitive to periods of prolonged drought. Where Swamp Paperbarks were once widely distributed and densely formed, they had been reduced to sparse isolated patches. Patches of dead, spindly trunks like those depicted in Figure 6.2 were evident across the property.

Beatrice and Jim were upset with the decline of Swamp Paperbark, as their property was one of the few in the area that had any habitat for the local Swamp Wallaby (*Wallabia bicolor*) population. Wallabies rely on the shelter that dense patches of Swamp Paperbark provide, while also utilising the foliage as a food source. As a result, Beatrice and Jim were fencing small sections of their remnant bushland, which was intended to allow the remaining drought-stressed Swamp Paperbark to slowly regenerate without being grazed by rabbits. While maintaining habitat potential for local fauna and returning the property to its 'original' state (i.e. its state prior to clearing for farming) were strong drivers

Figure 6.2 The dieback of Swamp Paperbark (*Melaleuca ericifolia*) in the foreground shows how the bush has changed through a long period of drought.

for attempting this restoration, Jim felt that 'mother nature has been against the (Paperbarks)', and that their efforts may be futile. It was clear during my visit that the fencing was having limited impact, despite recent rains, with few of the stressed shrubs recovering and little evidence of new recruits coming up through the soil. Beatrice and Jim posited that the lack of new seedlings might be a result of past vegetation clearing and cattle grazing on their property that had exhausted the seed bank in the soil.

The Swamp Paperbarks on Alex and Simone's property had also suffered during the drought, with a number of dieback patches like those seen in Figure 6.2. However, in contrast to the previous example, the loss of middlestorey species during the drought was not a concern for Alex and Simone, as the recent break in the drought had triggered strong growth of Sweet Bursaria (*Bursaria spinosa*) at a spot where a stand of Swamp Paperbark had previously died off (Figure 6.3). Despite not knowing the newly colonising species by name, Alex pointed to this as an example of the normal boom-and-bust cycles of the Australian bush. As Alex

Figure 6.3 In stark contrast to Figure 6.2, another precolonial benchmark species in Sweet Bursaria (*Bursaria spinosa*) has quickly established after the dieback of Swamp Paperbark, which occupied this spot only a few years prior.

and Simone had observed strong recruitment and regeneration of their bushland after periods of drought, the loss of Swamp Paperbark was just another example of the bush being a constant 'moving picture' (Simone).

Having experienced different ecological responses to the drought, Beatrice and Jim had developed an active disposition towards management practice, while Alex and Simone had taken on a passive, hands-off disposition (Erickson 2002; Trigger *et al.* 2010). The nature and extent to which landscape histories shaped the way Swamp Paperbarks responded to drought was complex and likely bound up with other dimensions, like differing soil types and surrounding ecologies. Yet, it is noteworthy that Beatrice and Jim sought to look backwards for their cues to management, to an ecology that persisted prior to the changes they had observed through the drought. Attempts to resurrect the ecology that they first encountered upon inmigration are perhaps not surprising when reflecting on the strength of redemptive conservation aspirations that many participants brought with them to their properties (Table 6.1). In contrast, the regeneration observed by Alex and Simone saw a greater willingness to be passive observers of their property's ecology, helping to guide a hands-off management disposition that was more open to surprise and uncertainty.

The potential for the lived experience of ecological change to shape ideas about ecological function and appropriate management has been well documented (Head and Muir 2006; Trigger *et al.* 2008; Cooke and Lane 2015a). However, in the instance above, we can see how lived experience and immersion *in* ecological change drove efforts to regenerate a drought-sensitive species in the face of ongoing climate change. These types of management interventions raise questions about the governance challenges of encouraging reflexive environmental management centred around robust, novel ecosystems (Higgs 2012; Young 2014). In the select examples here we see the potential for people–plant relations to generate strong dispositions for management and notions of 'what belongs' in the landscape. Combined with the structuring dimension of embodied landscape histories, the ongoing immersion of people in the landscape raises challenges for encouraging a transition away from precolonial ecological benchmarks for environmental management in the Anthropocene.

Insights for environmental management in the Anthropocene?

Rural-amenity landscapes provide a unique space in which to explore and progress new forms of environmental management due to the diverse landscape legacies, limited reliance on the land for productive output and active processes of landscape change that are already in the making. They are fertile ground for 'experimental' environmental management that embraces ecological function and process (Lorimer and Driessen 2014). Given the potential complexities for such a project raised by the examples above, what formative insights might we be able to draw that can contribute to the: (1) theory, (2) policy and (3) practice of environmental management in the Anthropocene?

Theory: landscape legacy as a trajectory for novel ecological futures

Decentring the ecological benchmarks that have traditionally underpinned environmental management means engaging with past ecologies in a way that recognises their coproduction and ongoing power. As Olwig notes, the landscape is a 'historical document containing evidence of a long process of interaction between society and its material environs' (2002, p. 226). When we recognise that past people–plant relations become embodied in the landscape and shape a trajectory for future ecologies, we are better placed to be attentive to the likely challenges and uncertainties for environmental management. The terminology of 'landscape legacy' may be instructive for communicating this notion (Cooke and Lane 2015b).

'Legacy' captures the idea of something being handed down from the past, with the potential to affect the future. As noted above, amenity migrants can hold strong redemptive aspirations that reflect a sense of wanting to be a 'steward' of the land, leaving it in a better condition than it was when they inherited it. While environmental stewardship comes in many forms, the desire to be active in creating and ultimately leaving a positive legacy is central (Worrell and Appleby 2000). Thus, we can think about landscape legacy as a trajectory of human and nonhuman interactions that sets a path for current and future environmental management (Cooke and Lane 2015b). These past legacies serve as the pathways that contemporary landholders are traversing in conjunction with nonhuman actors in the course of their management, and in the process, a course is being set for the future direction of ecologies that are already in the making. The entwining of stewardship aspirations with embodied landscape histories come together through environmental management to (re)produce ecologies, representing a moment where legacies are simultaneously bequeathed and inherited. Rather than harking back to precolonial ecologies for informing future environmental management, landscape legacy provides a heuristic that encourages *working with* the past. Through this lens, management becomes 'a process, not a destiny' (Ellis 2015, p. 27), where the need to be open to surprise and uncertainty in the policy and practice is easier to communicate.

Policy: framing rural-amenity ecologies on private land as 'inhabited'

The ongoing habitation of private land makes environmental management policy inherently different from traditional understandings of protected area conservation, particularly in the West. As Prévot-Julliard *et al.* (2011) note, it is vital that conservation policy combines an ecological emphasis with the need to reconcile people with the ecologies they inhabit. Ignoring people in this picture can mean we neglect to foster meaningful connections between people and plants, threatening the persistence of, and enthusiasm for, environmental stewardship. The narratives presented in this chapter and elsewhere (Cooke and Lane 2015a) suggest that the experience of dwelling in these ecologies can produce durable dispositions for management, making lived experience of the landscape a reference point for understandings of ecological function. As a consequence, perceptions of environmental change will impact on how policies or programs will be received and implemented.

Environmental management policy in rural-amenity landscapes could respond by framing ecologies as 'inhabited'. Inhabiting ecologies acknowledge more than a mental connection or 'knowledge of' environmental management; it encapsulates an experiential understanding of the landscape (Ingold 2011). As a consequence, external knowledge and policy prescriptions that may be promoting alternative management approaches require landholders to reinterpret their lived experience of the landscape, rather than simply changing their minds about ecological function. As has been shown, the changes to ecologies that are experienced by landholders can be keenly felt – as they were for Beatrice and Jim. Some landholders may pursue redemptive conservation aspirations with even more vigour as a consequence of observing the unanticipated (and potentially rapid) ecological changes to their surrounding environment. Policy efforts must be conscious of changing ecologies in the Anthropocene, as well as the *perception* of change as experienced by landholders. While amenity migrants may accept the emergence of novel ecologies following restoration efforts that diverge from their intentions, some may mourn the loss of plants that are no longer suitable to the shifting ecological niche they once inhabited.

An inhabited framing also helps to convey the idea that ecologies have been inhabited before, both pre- and postcolonial settlement. This positions rural-amenity migrants as part of an ongoing story of successional land use and habitation, rather than as stewards of benchmark ecological restoration opportunities that have presented themselves in light of the decline of productive agriculture. In this way, working with inhabited ecologies creates space for reflecting on landscape legacy and how current landscape inhabitants (both human and nonhuman) are navigating a trajectory of ecological change.

Practice: progressing carefully – what form of active intervention is desirable?

Acknowledging plant agency in the process of environmental management revealed the nuanced and unique ways that ecologies on private land are continually in the making. Of particular interest were the unanticipated ways in which plants 'acted back' to human intervention, which raises questions about the efficacy of a Western environmental management mentality that seeks to 'control' ecologies through management (Howitt and Suchet-Pearson 2006). In most cases, an active approach to environmental management is advocated in favour of passive approaches, as a means for managing threats like weed invasion and for restoring ecologies to highly modified landscapes (Erickson 2002; Klepeis *et al.* 2009; Abrams *et al.* 2012). However, participant narratives showed that passive management might not always result in problematic ecological responses, while active management is not always desirable for responding to anthropogenic change. Removing plants considered to be weeds may indeed remove some of the only complex floristic assemblages that are capable of providing habitat (Hobbs *et al.* 2006). Moreover, landholder perception of plants that 'act' like weeds can see them removed from the landscape (see Lauren), generating changes to ecologies that landholders ultimately consider to be undesirable.

This is not to suggest that environmental management should not be active in shaping future ecologies. Indeed, exploring ways to usefully and productively navigate environmental management is the core tenet of this chapter. However, it does suggest that research and practice adopt a critical perspective of active management that recognises how embodied landscape histories can set a trajectory for undesirable or unanticipated ecological assemblages if we do not progress carefully. This critical perspective will need to begin with a sensitivity to how the diversity and complexity of ecologies can be enhanced in a given place, in light of the structuring influence of landscape legacy. This means working with ecologies that are already given – building from existing species inter-relationships to improve ecological function and process, rather than using species origin as the starting point for active management interventions. There may already be an appetite for 'progressing carefully' with management in rural-amenity landscapes, given that half of the participants in this case study had pulled back from their early interventionist efforts following unanticipated responses.

Conclusion

In repositioning environmental management for the Anthropocene, we must be conscious that environmental management practice is a performative relationship that co-constitutes ecologies over time. In this sense, management practice is both a form of lived experience of the world that generates ideas about how it functions and what belongs (or does not belong) and a means for accessing and engaging with past human–environment relations embodied in those landscapes. Perceptions of what constitutes appropriate environmental management and ecological function are bound up in the bodily experience of plants, soil and other nonhuman elements that are themselves products of past performative relationships. Through inhabiting the landscape with plants over time, landholders are part of an ongoing conversation (Waage 2010) – a conversation that is picked up from the story of embodied histories present in the landscape and projected forward through the reshaping of ecologies.

As a result of experiencing embodied histories, environmental management in the Anthropocene may be complicated by the way people interpret ecological change, especially the loss of plants. In some instances, however, failed efforts to bring back precolonial benchmark ecologies may provide an opening for fostering reflexive environmental management. The management practices of rural-amenity migrants suggest the notion of 'landscape legacy' could be valuable for capturing people's dynamic participation in the transition of ecological histories into ecological futures. Landscape legacy encourages us to look at landscape histories as an inheritance rather than a benchmark to which we must return (Cooke and Lane 2015b). Policy aimed at rural-amenity migrants on private land can further disrupt ecological benchmark mentalities by positioning ecologies as inhabited. An inhabited perspective of ecologies draws much-needed attention to how perceptions of the environments we occupy will frame the way people respond to the policies and programs aimed at them. Finally, the people–plants relations that occurred through the management practices explored here emphasise the need for a critical perspective on *how* we promote active intervention in ecologies. The indeterminate ways in which ecologies respond to the acting of people in the landscape suggest the need to progress our management efforts carefully, with an emphasis on ecological form, function and process, rather than species providence. Continuing to interrogate environmental management practice at the fine-grained scale will be vital for uncovering ways of shifting our objectives from ecological redemption to reflexive environmental management in the Anthropocene.

References

Abrams, J., Gill, N., Gosnell, H. and Klepeis, P. 2012. Re-creating the rural, reconstructing nature: An international literature review of the environmental implications of amenity migration. *Conservation and Society*, 10(3), 270.

Argent, N., Tonts, M., Jones, R. and Holmes, J. 2010. Amenity-led migration in rural Australia: A new driver of local demographic and environmental change, in *Demographic Change in Australia's Rural Landscapes*, edited by G.W. Luck, R. Black and D. Race. Dordrecht, The Netherlands: Springer, 23–44.

Buxton, M., Tieman, G., Bekessy, S., Budge, T., Mercer, D., Coote, M. *et al.* 2006. *Change and Continuity in Peri-Urban Australia, State of the Peri-Urban Regions: A Review of the Literature Monograph 1.* Melbourne, Australia: RMIT University.

Castree, N. 2014. The Anthropocene and geography I: The back story. *Geography Compass*, 8(7), 436–449.

Cooke, B. and Lane, R. 2015a. How do amenity migrants learn to be environmental stewards of rural landscapes? *Landscape and Urban Planning*, 134, 43–52.

Cooke, B. and Lane, R. 2015b. Re-thinking rural-amenity ecologies for environmental management in the Anthropocene. *Geoforum*, 65, 232–242.

Curry, G.N., Koczberski, G. and Selwood, J. 2001. Cashing out, cashing in: Rural change on the south coast of Western Australia. *Australian Geographer*, 32(1), 109–124.

DEPI. 2007. Bracken fern poisoning of cattle. Accessed May 17 2015. Available at http://www.depi.vic.gov.au/agriculture-and-food/pests-diseases-and-weeds/animal-diseases/beef-and-dairy-cows/bracken-fern-poisoning-of-cattle.

Doody, B.J., Perkins, H.C., Sullivan, J.J., Meurk, C.D. and Stewart, G.H. 2014. Performing weeds: Gardening, plant agencies and urban plant conservation. *Geoforum*, 56, 124–136.

Ellis, E. 2015. Ecology in an Anthropogenic biosphere. *Ecological Monographs*, 85(3), 287–331.

Ellis, E.C., Antill, E.C. and Kreft, H. 2012. All is not loss: Plant biodiversity in the Anthropocene. *PloS One*, 7(1), e30535.

Erickson, D. 2002. Woodlots in the rural landscape: Landowner motivations and management attitudes in a Michigan (USA) case study. *Landscape and Urban Planning*, 58(2–4), 101–112.

Gill, N., Klepeis, P. and Chisholm, L. 2010. Stewardship among lifestyle oriented rural landowners. *Journal of Environmental Planning and Management*, 53(3), 317–334.

Gosnell, H. 2011. Amenity migration: Diverse conceptualizations of drivers, socioeconomic dimensions, and emerging challenges. *GeoJournal*, 76(4), 303–322.

Head, L. and Muir, P. 2006. Suburban life and the boundaries of nature: Resilience and rupture in Australian backyard gardens. *Transactions of the Institute of British Geographers*, 31, 505–524.

Head, L., Atchison, J. and Phillips, C. 2014. The distinctive capacities of plants: Re-thinking difference via invasive species. *Transactions of the Institute of British Geographers*, 40(3), 399–413.

Higgs, E. 2012. Changing nature: Novel ecosystems, intervention, and knowing when to step back, in *Sustainability Science: The Emerging Paradigm and the Urban Environment*, edited by M.P. Weinstein and R.E. Turner. New York, NY: Springer, 383–398.

Hobbs, R., Arico, S. and Aronson, J. 2006. Novel ecosystems: Theoretical and management aspects of the new ecological world order. *Global Ecology and Biogeography*, 15, 1–7.

Holmes, J. 2006. Impulses towards a multifunctional transition in rural Australia: Gaps in the research agenda. *Journal of Rural Studies*, 22(2), 142–160.

Howitt, R. and Suchet-Pearson, S. 2006. Rethinking the building blocks: Ontological pluralism and the idea of 'management'. *Geografiska Annaler, Series B: Human Geography*, 88(3), 323–335.

Ingold, T. 2011. *Being Alive: Essays on Movement, Knowledge and Description*. London, England: Routledge.

Kendra, A. and Hull, R.B. 2005. Motivations and behaviors of new forest owners in Virginia. *Forest Science*, 51(2), 142–154.

Klepeis, P., Gill, N. and Chisholm, L. 2009. Emerging amenity landscapes: Invasive weeds and land subdivision in rural Australia. *Land Use Policy*, 26(2), 380–392.

Lien, M.E. and Davison, A. 2010. Roots, rupture and remembrance: The Tasmanian lives of the Monterey Pine. *Journal of Material Culture*, 15(2), 233–253.

Lorimer, J. and Driessen, C. 2014. Wild experiments at the Oostvaardersplassen: Rethinking environmentalism in the Anthropocene. *Transactions of the Institute of British Geographers*, 39(2), 169–181.

Mastnak, T., Elyachar, J. and Boellstorff, T. 2014. Botanical decolonization: Rethinking native plants. *Environment and Planning D: Society and Space*, 32(2), 363–380.

Mendham, E. and Curtis, A. 2010. Taking over the reins: Trends and impacts of changes in rural property ownership. *Society & Natural Resources*, 23(7), 653–668.

Olwig, K. 2002. *Landscape, Nature, and the Body Politic: From Britain's Renaissance to America's New World*. Madison, WI: University of Wisconsin Press.

Power, E.R. 2005. Human–nature relations in suburban gardens. *Australian Geographer*, 36(1), 39–53.

Prévot-Julliard, A.-C., Clavel, J., Teillac-Deschamps, P. and Julliard, R. 2011. The need for flexibility in conservation practices: Exotic species as an example. *Environmental Management*, 47(3), 315–321.

Strang, V. 2010. Mapping histories: Cultural landscapes and walkabout methods, in *Environmental Social Science: Methods and Research Design*, edited by I. Vaccaro, E. Smith and S. Aswani. Cambridge, England: Cambridge University Press, 132–156.

Tilley, C. 2004. Mind and body in landscape research. *Cambridge Archaeological Journal*, 14(1), 77–80.

Trigger, D., Mulcock, J., Gaynor, A. and Toussaint, Y. 2008. Ecological restoration, cultural preferences and the negotiation of 'nativeness' in Australia. *Geoforum*, 39(3), 1273–1283.

Trigger, D., Toussaint, Y. and Mulcock, J. 2010. Ecological restoration in Australia: Environmental discourses, landscape ideals, and the significance of human agency. *Society & Natural Resources*, 23(11), 1060–1074.

Van Auken, P.M. 2010. Seeing, not participating: Viewscape fetishism in American and Norwegian rural amenity areas. *Human Ecology*, 38(4), 521–537.

Waage, E.R.H. 2010. Landscape as conversation, in *Conversation with Landscape*, edited by K. Benediktsson and K.A. Lund. Farnham, England: Ashgate, 45–59.

Worrell, R. and Appleby, M.C. 2000. Stewardship of natural resources: Definition, ethical and practical aspects. *Journal of Agricultural and Environmental Ethics*, 12, 263–277.

Young, K.R. 2014. Biogeography of the Anthropocene: Novel species assemblages. *Progress in Physical Geography*, 38(5), 664–673.

7 Wild tradition

Hunting and nature in regional Sweden and Australia

Michael Adams

It is dusk and I am hunting for rabbits. It is very cold. I walk slowly across the field, and through the scattered trees, grey kangaroos move off as I come close. As the darkness deepens, so does the cold. I have to keep swapping the rifle between my hands, to keep them both warm and functioning. I am very focussed, paying detailed attention to what I can see and hear and feel, and time moves mostly slowly, occasionally fast when there is a flash of movement and my adrenaline spikes. I see many kangaroos, a lone wombat, and startle some birds, but no rabbits. After maybe an hour it is too cold, and I head back to the truck, navigating carefully in the dark. I have not fired a shot, but I have watched Venus and Jupiter grow bright in the western sky and learnt that there are not always rabbits in this field.

Figure 7.1 Rusa deer, Australia.

I return a week later. This time I find a young, naïve rabbit, that does not know to run. But we do not shoot the young ones, there is not enough meat, so I just watch. Then, in a different place, I find an old, experienced rabbit – this one runs immediately without stopping (that is how it got to grow old) – so again I do not fire. My son hunts the next night and encounters what we assume are the same two rabbits, with the same result. We talk about how there seem to be so few rabbits this season that we are hunting specific individuals across a wide expanse of landscape.

Two weeks later I find the older rabbit again, and this time I shoot it. I am walking in the moonlight in the same place, just after sunset, and see the rabbit. It runs to cover in low shrubs and I circle up hill, not taking my eyes off the bushes. The rabbit and I see each other at the same time and it bounds off across the open field to the next cover. This happens twice more. The last time, I see the rabbit and have a clear sightline, drop to one knee and fire. One hour later, it is simmering in a pot on the woodstove. We own rural mountain country in southeast Australia, where we hunt for food and skins. We hunt maybe 50 days a year. We drink the rainwater that falls on that land, catch fish in the river and occasionally harvest wild plants. Wild food is part of our everyday diet, something very unusual for most Australians now.

Introduction

In this chapter I discuss human interaction with 'wild' landscapes, plants, fungi and animals in regional areas of Australia and Sweden. The chapter uses empirical and ethnographic data to connect with discussions about wild food, nature and belonging in two particular regions of Sweden and Australia. The role of first-hand encounter – embodied relationships to the knowledge of nature – and how this interacts with concepts of time underpins my examination of changing traditions. Similarly, the role of food cultures as an agent of transformation in national and cultural identity informs my discussion of belonging. As demonstrated in this chapter, wild is a challenging concept, subject to much recent discussion, and with different contexts in Sweden and Australia. Rabbits (*Oryctolagus cuniculus*) are an introduced species in Australia but live wild in the landscape. This same landscape also has free-ranging camels, horses and dogs, descendants of once companion animals. These are sometimes killed ('culled') as being out of place, not native. Some native species are also periodically culled (e.g. kangaroos in the nation's capital, Canberra; Vincent 2015). Where I live it is not legal to hunt native animals for food, but there are several introduced species we can hunt. In Sweden, hunting focusses on native species, including some that have reestablished after local extirpation (e.g. *vildsvin* or wild boar *Sus scrofa*). But many game species in Sweden are highly managed, so while they may live wild, their population dynamics and habitats are subject to extensive human manipulation.

As my parent's generation hunted in India, both for food (mostly mountain birds by my mother) and in situations of human–carnivore conflict (tigers and leopards by my father), I could argue my hunting is part of a family tradition. Tradition is the second part of my title and often invoked in arguments about hunting. I briefly review some historic antecedents of contemporary hunting and wild harvest in each country as context for further discussion of hunting traditions, including beliefs and behaviour.

Sweden and Australia have long histories of hunting. For most of Australia's human history, including early colonial settlement, acquiring wild food from the sea and the land formed the human diet, and hunting was a normal part of activity and foodways. Indigenous occupation for about 50,000 years was, and in some places continues to be, based on hunting, fishing and accessing wild plants. Modern non-Indigenous hunting traditions continue in the present, but are marginalised, controversial and contested. Northern Sweden has evidence of human occupation for about 10,000 years, based initially on Sami communities hunting reindeer and moose (Bergman et al 2004). In modern Sweden, in part because of the historic traditions of *allemansrätten* (the right of public access) and the more recent *friluftsliv* (literally, the free air life), the use of wild food including hunting in contemporary times is much more normalised. In both countries the percentages of the population that are hunters are relatively similar, between 2 and 5 per cent.

Reflecting these long traditions, the Jämtland region, a province geographically in mid-Sweden but considered significantly 'northern', and the focus of my Swedish study is unique for the intensity of archaeological sites associated with hunting moose. Five-thousand-year-old petroglyphs illustrating moose and many hundreds of trapping pits are found throughout the region. In Jämtland I interviewed active hunters and their families contacted through hunters already known to my wife's Swedish family. In Sweden almost all my discussions with hunters have been in their homes or outdoors, over shared meals with their families, usually of wild foods they have hunted or foraged.

In Australia I surveyed hunters in my home region, the Illawarra on the southeast coast in the state of New South Wales, and then met with and interviewed active hunters both in large meetings and individually, following up from the surveys. So far my discussions with hunters in the Illawarra have been in meeting rooms or other neutral locations, cafes, classrooms, outside, and I have not yet shared meals. These different interactions reflect both differences in methodologies (I had stronger personal connections with Swedish hunters through family relationships and existing friendships) and the varying levels of social acceptance of hunting in these countries. Most European countries including Sweden maintain detailed and accessible hunting statistics, including number of hunters and quantification of the bag taken (number of animals hunted). In Australia, that data is non-existent or very uneven, with no clear statistics either nationally or for the state of New South Wales for either hunter numbers or number and species of animals hunted. My survey in the Illawarra, while focussing on deer, also requested information on other species and quantities hunted.

Sweden: hunting in Jämtland

Recent studies have provided summaries of wild food use in Europe and Sweden, reviewing geographic patterns, intensity, and species and quantities of wild food accessed in twenty-first-century Europe. For Europe overall, Schulp *et al.* (2014) estimate that 14 per cent of the population (65 million people) collect wild food occasionally, and at least 100 million consume wild food. In 2005, 26 million kilograms of hunted meat was marketed. This potentially represents only a fraction of the actual total game meat accessed, as much does not reach the monetary economy but is distributed and consumed through hunter networks and families. In addition to game meat, both mushrooms and berries are harvested in large quantities, including tens of millions of kilograms of berries in Sweden and very significant quantities of mushrooms. Several species of berries and fungi are collected, including *Lingon* (*Vaccinium vitis-idaea*) and *Blåbär* (*Vaccinium myrtillus*), *Kantarell* (*Cantharellus cibarius*) and *Trattkantarell* (*Craterellus tubaeformis*).

While acknowledging that wild harvest represents a small fraction of total food consumption, Schulp *et al.* (2014) argue that the health and cultural benefits are high, and that for some rural or poor communities in parts of Europe outside Sweden, the nutritional value is significant. Łuczaj *et al.* (2012), focussing just on wild plant use in Europe, analyse trajectories of change from 'poverty food' through to the availability of cheap sugar in the twentieth century, enabling very significant increases in berry harvest and preservation. They document negative changes, including increasing urbanisation leading to loss of plant knowledge and contact with nature. Focussing then on the present, they examine a recent spike in attention to wild food plants deriving from perceived health benefits as well as the rise of regional cuisines, agritourism and avant-garde cuisine.

Both Svanberg (2012) and Stryamets *et al.* (2015) argue that few wild plants or fungi were regularly used in preindustrial Sweden. The availability of sugar led to a rapid increase in berry harvest, particularly among urban people, with nearly 60 per cent of the population picking berries. Mushrooms, initially despised by rural people, became popular from the nineteenth century, with another spike in the 1970s. Wild plants and fungi are also harvested commercially, with a recent rise in the involvement of migrant labour.

Hunting by Sami, particularly reindeer-herding communities, while having ancient roots, is increasingly politicised in modern Sweden. In 1992 the bill to create the Swedish Sami Parliament included changes in legislation that removed exclusive small game hunting rights from Sami in north and mid-Sweden. Sami perceived that the new Sami Parliament was being paid for by the loss of their unique hunting and fishing rights. Green (2009) argues that this continues to be a 'burning issue' for many Sami, as it has allowed many more hunters to access game previously only available to Sami.

Today, *Svenska Jägareförbundet* (the Swedish Association for Hunting and Wildlife Management) is the primary hunting organisation and maintains extensive data on Swedish hunting. There are around 300,000 hunters in Sweden, a little over 3 per cent of the population, or 5 per cent of the population between 18 and 75 years old (Boman *et al.* 2011). Most of those hunters are men, but a grow-ing number of women are gaining hunting qualifications and registering to hunt. Recent statistics suggest 15,000 women have hunting permits, and 25 per cent of those undertaking the 'hunting exam' are women. Promoting involvement of women is one of the initiatives to recruit more hunters to what is a declining and ageing hunter population. Public support for hunting is strong in Sweden (Ljung *et al.* 2012), so these declining numbers reflect increasing urbanisation and an overall ageing of the population. Many older hunters I spoke with, and the adult children of hunters speaking of their fathers, indicated that in many mid and northern rural regions 30 years ago, almost all protein was wild-hunted or fished.

Several species are commonly hunted in Sweden, with 16 mammals includ-ing three species of deer (*Capreolus capreolus, Dama dama, Cervus elaphus*) wild boar (*Sus scrofa*), hare (*Lepus timidus*), beaver (*Castor fiber*) and preda-tors including bears and foxes. Forty species of birds may be hunted, including capercaillie (*Tetrao urogallus*), grouse, pheasant and many water birds. In mid- and north Sweden though, the primary prey is moose (*älg* in Swedish, *Alces alces*), with nearly 100,000 moose shot annually across the nation, a third of the total population. In the last few years, nearly as many wild boar have also been shot, with both their numbers and distribution expanding rapidly. For all species hunted, the total annual bag is more than 600,000 animals.

Sweden has a large and important forestry industry, and the large moose popu-lation has developed since the transition from selective logging to clear-cutting in timber production from the 1960s. Regenerating timber clear-cuts provided moose with excellent food sources, simultaneously creating an incentive for the forestry industry to promote moose hunting to limit damage to pine seedlings (Hammarström 2004). Currently game meat harvest in Sweden is estimated at 16 million kilograms annually, of which 11 million kilograms comes from moose. In Jämtland 13,251 moose were reportedly shot in the 2012 season (Wikner 2015). Likely more than this were killed and came into the social networks of food exchange via recovery of road-killed/injured moose, which is a responsibility of local hunting teams.

In Jämtland and other parts of Sweden, hunting is positioned in a spectrum of regional food practices (discussed further in a later section); it is also posi-tioned in a broader spectrum of outdoor recreation activities. While only 3 per cent of Swedes hunt, nearly 90 per cent of Swedes hike in forests or nature at least six times per year, and nearly 40 per cent cross-country ski (Romild *et al.* 2011). A comparable number of nearly 40 per cent fish at least once per year (Romild *et al.* 2011), and around 23 per cent regularly participate in fishing (Arlinghaus *et al.* 2015). This is closely linked to the Nordic idea of *friluftsliv*, 'the free air life', stemming initially from rural traditions and then made more explicit with a nineteenth-century focus on health, romanticism and nationalism, and itself enabled by the open-access tradition of *allemansrätten*. While *friluftsliv* is often positioned as a normative identifier of Norwegian and

Swedish culture and identity, it has historically been highly gendered (Gurholt 2008). The earliest published use of the term is from the Norwegian playright Henrik Ibsen in 1859, who used it in a masculinised hunting context. From the 1890s, middle-class Norwegian women began to use the concept as an expression and symbol of emancipation, and there continues to be active contestation of the dominance of men (Gurholt 2008). Publications currently used to promote Jämtland feature outdoor women very prominently, including several as hunters (Wikner 2014, 2015).

Hunting is highly regulated in nearly all aspects: every species has a particular open season and bag limit, a compulsory hunting exam is mandated by the state, and hunting proficiency tests are usually required annually by local hunting clubs. This level of prescription, described to me by an American living in Sweden (DL interview 2015) as 'insane regulation', is one reason for Sweden's broad public acceptance of hunting – there is a strong assumption that hunters are careful and ethical, and hunting is a safe activity. Ljung *et al.* (2012) found an acceptance rate for hunting of 80 per cent of the non-hunting population, supported by consumption of game meat and social relationships between hunters and non-hunters.

Figure 7.2 Moose hunt, Sweden.

Credit: Ingrid Johansson

In much of Sweden, the hunt is a social event. Moose hunters work in groups and share the bag irrespective of who actually took the successful shots, so you can come home with moose meat from a hunt even if you did not ever see a moose. While it is highly social, and while some have argued the social is at least as important as the hunt itself (Hammarström 2004), it does not regularly include the sort of formalised hunting rituals encountered in some other parts of Europe. One interviewee suggested that for many Swedes, the moose hunt is still closer to the *allmoge-jakt*, or peasant hunt, where 'the forest and the wild-life in it always played an important role in contributing to the household' (LM interview 2015). Gunnarsdotter (2007) describes local hunters' environmental relationships as an 'enchantment with the forests' (p. 190) and argues that they 'perceive the meaning of hunting as a wholeness or fusion of hunter-forest-game-place-history' (p. 183). While she also discusses the rise of hunting tour-ism, and how this might impact these relationships with the monetised value for hunting changing intrinsic values to instrumental values, she suggests that these are often different groups of people. Tourist hunters do not know the landscapes or the animals and have quite different relationships with the forests. This is in part temporal: locals have long relationships with place, tourist hunters are brief visitors who need to be guided.

One of the hunters I interviewed, a retired dentist in his seventies, told me that he ate only wild-caught meat and fish in his household. He is a very skilled hunter, with deep knowledge of the animals and habitats of his hunting region, as well as much experience in other places. His knowledge of his regular hunting terrain was reflected in his observation that increasing numbers of brown bears had in 2014 killed every moose calf born that year. He hunted to his quota maximums every year and also indulged in some slightly illegal fishing – these activities filled his family freezer with enough protein for each year. Describing the moose hunt, he said he was successful because he stayed completely motionless, for up to 6 hours at a time in the hunting stand – 'the *älg* are watching you all the time, so you don't drink, you don't urinate, you don't look at your phone'. This is both a demonstration of strength – while not young he can sustain this kind of attention for hours – and an indication of how there is a different kind of time in these circumstances – he is immersed in the enchantment of the forest (DH interview 2015).

While this man, retired and living close to hunting terrain, could hunt more than 50 days annually, most Swedish hunters spend less time. The average for moose hunters is 9 days annually (Wikner 2015), and hunters in general average 20 days annually (Boman *et al.* 2011), but even these hunters try to eat wild meat regularly. All the hunters I interviewed focussed on the food aspect of the hunt. At every meal I shared outside my own immediate family, I ate wild-caught fish, meat, berries and mushrooms, although this may reflect a conscious effort by these hunters to demonstrate these foods to me as a researcher and visitor. Several people said they tried to harvest enough berries so they could eat them every day throughout the year, and large chest freezers stocked with moose or roe deer were common. Some hunters in small villages combined to share large commercial-scale frozen storage facilities.

The sense of the large landscape of forests, lakes, rivers and swamps as a source for provisioning was very strong. Across Jämtland, farmlands are being abandoned, and it is not at all unusual to come across fully furnished, quite elaborate residences collapsing in decay because no buyers can be found for them. With the withdrawal of people from these landscapes, game animal populations have increased. Recent years have seen bag limits and rules increased for several species, including acknowledgement that expanding wild boar populations can no longer be contained (DH interview 2015). This is another temporal aspect: the long processes of fluctuation of human and animal populations in the same place, held in memory by long-term residents, and starkly evident to visitors such as myself.

While I experienced much evidence of a vibrant hunting culture in Jämtland, I also spoke with people who found difficulty accessing hunting opportunities (SW interview 2015). These were more recent arrivals, and their desire for wild food was addressed through their own foraging practices, through hunted meat being shared through workplace relationships, and by purchasing wild meat in local stores.

Australia: hunting in the Illawarra

There are no reliable statistics on hunting in Australia. Estimates of active hunters nationally range from 4 per cent (Bauer and Giles 2002) to 1.5 per cent (Finch *et al.* 2014). This quite large variation is based on a number of factors: there are no centrally collected data; many Australian hunters deliberately avoid organised and regulated hunting activities; and some hunters use bows, dogs and knives, rather than firearms, and there is little regulation and no statistics on these hunters.

Many Australian Indigenous people continue to hunt, ranging from communities that continue to live on what are often considered 'remote' homelands in northern and central Australia (somewhat analogous to northern Sweden); to urban members of communities who maintain hunting traditions both by travelling and by local peri-urban hunting. Some, at least, of these people hunt 'illegally', arguing that they have millennia of unbroken tradition of hunting that is not responsible to 'whitefella' law. This continuity of hunting may show marked changes in technology but a consistency in purpose: ancient practices of sharing food continue, and a visceral desire for wild, native meat is expressed. There are numerous studies of remote Indigenous hunting (e.g. Altman 1987, 2012), but almost none focussing on the 60 per cent of the Indigenous population that live in large rural and urban centres.

In New South Wales, it is not legal to 'recreationally' hunt kangaroos and wallabies for meat. There is a commercial kangaroo harvest that is geographically restricted to the western part of the state, with most of the product from this being exported. On our mountain place, there are healthy populations of at least three species of macropods: eastern grey kangaroos (*Macropus giganteus*), red-necked wallabies (*Macropus rufogriseus*) and swamp wallabies (*Wallabia bicolor*). Although I hold firearms and hunting licences, I am not permitted to shoot any of these animals for food. I could get a licence to 'cull' them if I argued that they

Figure 7.3 Rabbit hunting, Australia.

were a problem for my agricultural activities, but once shot, I cannot use the meat or hides. If I want to eat kangaroo (which my family often does), we need to drive an hour to the supermarket to purchase wild-shot kangaroo that originated at least hundreds of kilometres away.

As no native animals can be hunted in New South Wales, hunting focusses on a suite of introduced animals, including several species of deer. My research in the Illawarra region focussed primarily on deer hunting, with the most abundant species being Javan rusa deer (*Rusa timorensis*). The Illawarra is likely broadly representative of both the variation and patterns of hunting in Australia. The topography of a steep coastal escarpment unsuitable for development creates significant habitat that has been rapidly occupied by several deer species, initially escaped from deer farms.

Data from the survey (conducted in late 2014) indicated that at least 500–600 rusa deer were successfully hunted annually in the region, constituting around 15,000 kg of useable venison. From the survey responses, 80 per cent of hunters harvested around 150 kg, and a small subset of very active hunters reported between 500 and 2,500 kg of deer meat hunted annually. A significant number of hunters ate wild meat at home every week, while others ate wild meat monthly or less. In strong contrast with Sweden, almost no hunters surveyed or interviewed indicated that they also harvested wild plants.

While there is certainly an interest in wild plant harvest in Australia, in this region those who engage in this practice may morally distance themselves from hunters. There are numerous books on wild plant use by Australians, most of which mention hunting only in the context of Indigenous or colonial practice (Isaacs 1987; Cherikoff and Isaacs 1989; Low 1989). Recent active websites on the topic are similar (e.g. Bonetto n.d.). There are no statistics on wild plant harvest, but the number of people participating is likely to be very small. At least 17 per cent of Australians are regular fishers, but unlike Sweden, participation in outdoor activities in general is not particularly high, with only 2.5 per cent regularly hiking (ABS 2015).

The Illawarra and South Coast regions of New South Wales have relatively recent agritourism and 'food and wine' tourism activities, following the growing significance of tourism in the region. In parallel, local authorities recently prepared an *Illawarra Regional Food Strategy* (Campbell 2013). While comprehensive in many respects and including quantification of local agricultural output, it entirely ignores food provisioning by recreational fishing and hunting. Fishing in particular supplies thousands of kilograms of protein to Illawarra households: the recreational fishing catch for at least six species is significantly greater than the commercial catch, and the Illawarra is the fourth most active fishing community in the state (DPI 2001). The hunting survey indicated that thousands of kilograms of hunted venison also finds it ways on to Illawarra dining tables annually. Unlike Sweden, in New South Wales it is not legal to sell hunted meat commercially (except for the commercial kangaroo harvest), so this meat is distributed through hunter and family networks of exchange. From the survey, it is likely that several hundred Illawarra families benefit from this informal distribution. As in Sweden, many Australian hunters are active fishers also, but unlike Sweden, there is a common moral distinction made by many people between these roles. As one hunter said, 'if I take my son or daughter fishing, I'm the greatest dad in the world, if I take them hunting, I'm a barbarian' (TB interview 2015).

Hunting and national cuisine: Sweden and Australia

As some elements of the history described earlier demonstrate, tradition, while by definition referring to beliefs and practices transmitted from the past, can also be invented, and invented traditions are often common in modern national cultures. One of the most striking recent examples in Sweden is the development of what has been called the 'new Nordic kitchen'. In 2004 the 'New Nordic Cuisine Manifesto' (Meyer 2004) was published with a set of aims including: 'to express the purity, freshness, simplicity, and ethics' of the region; to promote animal welfare including 'in the wild'; and to support local self-sufficiency and regional sharing. In the decade since, the ideas and outcomes of the new Nordic cuisine have rapidly expanded. Initially led by restaurants and avant-garde cuisine (e.g. Noma in Denmark and Fäviken in Jämtland, Sweden), discussion of the concept has spread into popular culture and is actively supported by national and regional governments. While it includes individual hunting and foraging, there are also professional foragers who provide significant quantities to some markets (Larsen and Österlund-Pötzsch 2013). Jönsson maps the historic development of this movement and argues that 'reading a menu in a fine-dining restaurant in Sweden has gradually turned into a geography lesson' (2013, p. 60). This cuisine is both placed and temporal: particular foods are mapped to their origins and seasons; links to past traditions are identified and innovations are noted.

Sweden is one of the fastest urbanising nations in Europe. Jämtland both reflects and contradicts this: while its only city, Östersund, is growing and includes a third of the population, it has only 45,000 people and the province overall has a declining and ageing rural population; in addition, it is one of the least intensely settled regions in Sweden. Östersund also mirrors the changing patterns of wild food use. In 2011 it was declared a UNESCO 'City of Gastronomy' after a campaign led by Fia Gulliksson, a local food entrepreneur married to an active hunter. This designation has transformed the region. Building on what was already a strong national reputation for regional food specialities and outdoor activities, the Creative City designation has supported a significant expansion of city restaurants, speciality food outlets, agritourism and hunting tourism. It has been strongly supported by government, including a high-quality publication, *Jämtland Härjedalen Gastronomy* (Wikner 2014, 2015), and several websites (e.g. Hedberg n.d.). Throughout this promotion, hunting is integrated into the presentation of diverse local food systems.

That promotion is accurate, because hunting in much of Sweden is distinctive for its inclusion of broader approaches to accessing wild food. All the hunters I spoke with also harvested berries and mushrooms at least, often fished, and some harvested a wide range of products, food and otherwise from wild sources (these included firewood and construction timber; birchbark, pinecones and other materials for crafts; birch sap as a base for drinks; and medicinal plants). In Jämtland, hunting is seen in a continuum with ideas of other 'local food,' with a very particular focus on the unique qualities of the region. These qualities are expressed in plants and animals raised on farms and in plants and animals living wild in the matrix of forests, lakes, swamps and rivers. The Slow Food idea of 'terroir' is seen to manifest in both wild and cultivated species. This has both a long history and a recent revival in the context of local, healthy and environmentally sustainable food. Overlapping with the 'locavore' concept (Star 2010), the inclusion of hunted and fished food resources is seen to create additional benefits. A locavore hunter simultaneously engages in alternative food economies; achieves health benefits both from exercise and nutrition; increases well-being from contact with nature; and participates in local ecosystem management (Tidball *et al.* 2013).

In contrast to Sweden, most Australians today eat essentially the relatively limited suite of foods brought from Britain by the colonising First Fleet in 1788, despite living on a continent where Indigenous people have long demonstrated that there are thousands of flourishing edible species. The only clear overlap in species is marine and freshwater fish and invertebrates, where both colonial and contemporary use reflects the millennia of Indigenous harvest. The relatively short two centuries of occupation of Australia has uneven expression across the country. Many northern Indigenous communities demonstrate strong continuities of practice and immersion in their homelands, while in the Illawarra and other areas of the southeast, daily Indigenous customary practices persist, often invisibly, in the interstices of a transformed landscape.

The early colonial period in New South Wales (the first region colonised in the southeast of Australia) is interesting for a range of responses to wild food. Newling (2011) argues convincingly for what was initially a class-based response, where officers considered native animals and birds as game meats which were an established part of the British upper-class dining traditions. Convicts, soldiers and others accessed wild foods to supplement limited rations, but often saw them as inferior. This hunting also had to be clandestine, as all food acquired was legally required to be handed in for centralised distribution. Clandestine hunting was obviously a continuation of poaching, reflecting the old British division where hunting was reserved for the aristocracy, and hunting by others was poaching and punishable (this also has historic parallels with Sweden). Some plant foods were used that were not part of the Indigenous diet, as the colonisers brought boiling technology to render palatable new species, particularly plants used for tea substitutes. The other key point Newling makes is that wild food resources were essentially incompatible with a food system that was based on a 'consistent efficient and manageable food supply to support a permanent settlement' (2011, p. 40), that is, the food systems were enmeshed in existing labour systems and social structures not conducive to sustainably accessing wild sources.

As the Australian colonies expanded, engagement with hunting varied. James Boyce (2008) describes Tasmania, Australia's southern island state, as having a 'kangaroo economy' for the first 20 years of the colony, with the first generation of convict settlers adapting to an island environment rich in resources. From 1803, forester kangaroos (*Macropus giganteus*, known as eastern grey kangaroos on the mainland) were initially very plentiful and easily hunted with dogs and formed an almost complete replacement for imported meat and material for clothes and shoes. Living a 'pre-industrial' lifestyle, convict hunters learnt bush skills and lived off the land, endorsed by the colonial administration and supplying kangaroo meat as rations to the settlements. In Tasmania it continues to be legal to hunt several macropod species, including red-necked wallabies (*Macropus rufogriseus*, called Bennetts wallaby in Tasmania).

In contemporary Australia, native food sources are periodically explored as part of discussions around a national cuisine, but continue to be a very marginal part of popular diets. There is a boutique 'bush food' industry, with relatively tiny quantities of products from some local species available through web ordering and speciality stores. It is also possible to buy kangaroo meat in supermarkets, but very limited numbers of Australians regularly eat kangaroo (Waitt 2014).

While the deer survey indicated a significant amount of hunted venison coming into some Illawarra households, it would still constitute a relatively minor part of most people's food ingredients. Santich (2011) argues that while there was historically inclusion of native foods through to the nineteenth century, Australians have become progressively 'estranged from native ingredients.' Native foods feature in expensive restaurants, but are not regularly part of most Australians' food choices.

In Sweden the 'new Nordic cuisine' has revived discussion of these issues, and some local species are a regular part of popular diets. The berry *lingon* is widely available in supermarkets, sold internationally through Ikea, and often considered a staple of Nordic diets. But Sweden also demonstrates the wide gap between ideas of wild Nordic foods championed by expensive restaurants and the daily diets of Swedes in many southern and urban areas. Jämtland and Östersund are perhaps unique in the wide variety of wild, local food available and incorporated into local food choices as well as being featured in restaurants.

First-hand encounter

> Firsthand knowledge is enormously time consuming to acquire; with its dallying and lack of end points, it is also out of phase with the short-term demands of modern life. It teaches humility and fallibility, and so represents an antithesis to progress. It makes a stance of awe in the witness of natural process seem appropriate.
>
> (Lopez 2001)

The data from surveys and the hunter narratives tell several stories. One compelling story is about first-hand encounter – the immersed, embodied, visceral, temporal engagement between hunters and the environments and animals of their hunting practice. As Lopez indicates, first-hand knowledge, the product of first-hand encounter, is very time-consuming to acquire and teaches humility and fallibility.

Reflecting on my opening narrative, successfully hunting one rabbit took four nights and two people. That is clearly unsuccessful if evaluated on the basis of relative calories expended and acquired. But evaluated as learning, as acquiring knowledge, it is strongly effective. Intellectually we observe, share, discuss and analyse; viscerally, we listen, watch, shiver, sweat, feel our adrenaline and calm our breathing. Bodily learning takes place: our feet find paths across dark uneven ground, our eyes quickly find focus through the rifle sight, our breathing adjusts as we squeeze the trigger. We learn the individual habits and home of a particular animal in a particular place at a particular time and build knowledge of larger seasonal and diurnal time cycles and spatial patterns. A whole lot of our learning is probably not even conscious – muscle memory, patterns in the landscape, understanding our fickle mountain weather.

Bawaka *et al.* (2015) write of Indigenous women in Bawaka country in northern Australia: 'They pay close attention to what's happening around them and listen carefully to the messages that are sent out' and 'What humans do is attended to by many others'. While I do not necessarily want to draw a strong connection between hunters as a broad category and Indigenous peoples as a broad category, the idea of *attention* is very relevant. Hunters say they 'look for sign', and while that can be interpreted mundanely to mean tracks, droppings, antler rubs – the physical signs of animals' presence – it is used more broadly to indicate awareness of appropriate feeding and shelter habitat, impacts of weather conditions across the landscape, diurnal patterns of movement. These are all also deeply temporal: tracks might be minutes or days old, and good hunters can tell the difference; daily patterns will change depending on the season. This temporality is both within the hunter's skills and knowledge (in this weather, the sharp edge of that track means maybe an hour old or less) and within the hunter's memory (last year deer sheltered from south winds in the afternoon in that gully).

In *Meditations on Hunting* (first published in 1942) routinely referred to as the quintessential expression of the Western traditions of hunting, Ortega y Gasset says 'the hunter ... needs to prepare an attention which does not consist in riveting itself on the presumed but consists precisely in not assuming anything and in avoiding inattentiveness' (1995, p. 130). Similarly, it is very obvious that other beings in the landscape pay attention to what hunters, human and otherwise, are doing: the alarm calls and flights of birds; unidentified sound close by in thick bush; the focussed ears, noses and eyes of prey animals; all attest to the attention and agency of the Earth others participating in the hunt. Some of these may become the target of a hunter's shot; some may benefit from the remains of the hunted animal left in the forest; some may hone evasion skills. While Gasset argues for the attention of the predator, others have argued for the attention of the prey, acknowledging 'our vulnerability to nature's inhospitability' (King 2010). In both Sweden and Australia this applies: bears, crocodiles, snakes, difficult terrain and weather are contexts that do not guarantee human safety.

In modern societies, this kind of attention has eroded – increasingly knowledge of environments and interaction with nature is mediated by photographic or videographic image or text. This starts in childhood (Louv 2008) but continues into adulthood (Louv 2011). In many modern societies, back-country outdoors activity is declining, with ageing and urbanising populations and the rise of a risk society. Hunting can be seen as simultaneously pushing back against these declining connections, an anachronistic continuity of tradition from greying ranks of hunters, and stepping forward into revised ideas of locavore consumption and nature-embedded terroir, led by younger and often women hunters.

Hunting, nature and belonging

> The sense of place is also a critical component of the hunter-forest-game-place-history synthesis ... the place becomes a part of the hunt instead of an object for the hunt ... a well-known landscape communicates with us and makes us remember by addressing all our senses.
>
> (Gunnarsdotter 2007, p. 190)

Hunters in Jämtland are 'at home' in their nation and at home in the forest. There are old traditions, a strong history of rural and outdoors practicality and competence, and wide acceptance of eating wild foods. The new Nordic cuisine emphasises belonging; these are the foods that best flourish in Swedish landscapes, and harvesting, hunting and growing them, eating them fresh and clean, often outdoors, is emblematic of being Swedish. The actual practice of acquiring these foods is healthy activity, both physically and emotionally. The landscapes of the hunt are part of the rich and diverse landscapes of productivity in Jämtland: agricultural fields, productive forests, streams and lakes. While the experience of Sami hunters problematises this simple narrative, Sami people and politics are very marginalised in Sweden (compare Chapter 10 by Lawrence and Åhrén in this book).

In Australia, with its colonial history of invasion and dispossession, and an almost complete replacement of food sources from local to imported, the concept of belonging is often contested. For the non-Indigenous hunters I spoke with in the Illawarra however, it was not questioned. These people see their competence in the bush as evidence of their belonging. Their embodied skills and knowledge, their ethical practice, their relationships with the animals and habitats in which they hunt express the authenticity of their place. The undercurrent in their conversations is that it is the outcomes of their skills and knowledge – that is, the reactions of the other sentient and sapient beings in the bush – that legitimise them; their ability indicates their acceptance in the community of beings of the bush. They are aware that others in the bush are keenly and intelligently aware of their actions and their intent, and that they need to be serious, skilled and ethical if they are to be successful.

In the national consciousness though, Australian hunters are marginalised. A combination of the historic division between an inherited right to hunt and poaching; hunters' resistance to regulation and a 'socialised' hunt; and many hunters' political tendency to advocate the US model of gun (un)control – all frame hunting as a marginal and dangerous activity. They hunt subversively in the interstices of the landscape: sometimes on farms, legally and illegally; on residual lands abandoned by mining or agriculture and too degraded for national parks. They are discreet or secretive and highly defensive in hunter organisation publications and statements.

For both Swedish and Australian hunters, their belonging as hunters is also fundamentally through the fact that they hunt for food. The flesh of the animals they hunt becomes their own flesh. All the dimensions of the life of the animal they eat – the air it breathed, the water it drank, the food it ate, the muscles it built in its quotidian life – are incorporated into their own bodies and the bodies of those with whom they share the hunt.

This is an intense personal expression for many active hunters. It sits within and alongside broader national discussions of belonging that also co-opt food. The new Nordic cuisine movement does this explicitly in the construction of contemporary Nordic identity, with a 'natural' food culture as a bearer of social values, meaning and tradition. Australia's exploration of ideas of national cuisine has repeatedly included native animals and plant species, from colonial times to the present. While in both countries there continues to be a large gap between the public discourses of food culture and everyday eating practices, in Sweden wild food is far more commonly part of many everyday diets than in Australia.

Conclusions

Hunters kill. Like all living things, they sustain their lives through the deaths of others. Unlike many non-hunting people, they (mostly) acknowledge the deaths they cause. For hunters, what they eat is nutritionally significant (low fat, no hormones, high iron, and so on); it affects who they become, their strength and capacity to continue to hunt. But it is also significant in a much less rational context: in eating wild food they viscerally embody the wild: this is deliberate and conscious and paradoxically positions them with groups such as vegans or vegetarians who make very deliberate ethical choices about their food.

Situating this discussion in contemporary predictions of impending abrupt social and environmental change, hunters' embody their relationships to the knowledge of nature. Their engagement with their environments is built over long periods of physically strenuous practice and intense focus on the other species that share the landscape. They are a small percentage of the populations of Sweden and Australia, enacting what many consider anachronistic traditions in modern society. These traditions are not however linear: younger, sometimes women, hunters both revive old continuities and create new and different relationships. Changing attitudes to food champion wildness, regionality and freshness. In a context of declining engagement with nature and increasing commodification of experience, hunters have unique relationships. In Jämtland's declining rural populations their activities help maintain social networks, with other hunters and consumers of wild food, and also with the non-human others of the forest. In Illawarra, where hunters are more often alone, their sharing of wild meat connects outwards from their often intensely solo engagements with the other inhabitants of the bush.

The histories and geographies of each of these regions clearly shape attitudes and practices in hunting and in eating wild food. There are parallels and contrasts. These regions reflect the new but uneven global practices and principles of local, sustainable and organic, but are structured around very different national moralities and identities. Contemporary practices of hunting, fishing and foraging for wild foods are situated and defined by these places and times. While more accepted in Sweden than Australia, hunting is contested in both countries, but may offer new possibilities for innovative, hybrid, creative ways to reconsider how humans relate to the other beings on the planet, including the ones we eat.

Acknowledgements

Many people assisted with the research supporting this chapter. I would like to acknowledge and thank all those who agreed to interviews and surveys, and two outstanding research assistants at the University of Wollongong, Eli Taylor and Adrienne Corradini.

110   *Michael Adams*

References

ABS. 2015. *Participation in Sport and Recreation 2013–2014.* Canberra, Australia: ABS.

Altman, J. 1987. *Hunter-Gatherers Today: An Aboriginal Economy in North Australia.* Canberra, Australia: Aboriginal Studies Press.

Altman, J. 2012. Indigenous futures on country, in *People on Country: Vital Landscapes, Indigenous Futures,* edited by J. Altman and S. Kerins. Sydney, Australia: Federation Press, 213–231.

Arlinghaus, R., Tillner, R. and Bork, M. 2015. Explaining participation rates in recreational fishing across industrialised countries. *Fisheries Management and Ecology,* 22(1), 45–55.

Bawaka Country, Wright, S., Suchet-Pearson, S., Lloyd, K., Burarrwanga, L., Ganambarr, R., *et al.* 2015. Co-becoming Bawaka: Towards a relational understanding of place/space. *Progress in Human Geography.* doi:10.1177/0309132515589437.

Bauer, J.J. and Giles, J. 2002. *Recreational Hunting: An International Perspective.* Gold Coast, Australia: CRC for Sustainable Tourism.

Bergman, I., Olofsson, A., Hornberg, G., Zackrisson, O. and Hellberg, E. 2004. Deglaciation and colonization: Pioneer settlements in northern Fennoscandinavia. *Journal of World Prehistory,* 18, 155–177.

Boman, M., Mattsson, L., Ericsson, G. and Kriström B. 2011. Moose hunting values in Sweden now and two decades ago: The Swedish hunters revisited. *Environmental and Resource Economics,* 50(4), 515–530.

Bonetto, D. n.d. Accessed 19 August 2015. Available at http://www.diegobonetto.com/about/#bio.

Boyce, J. 2008. *Van Diemen's Land.* Collingwood, Victoria: Black Inc.

Campbell, R. 2013. *Illawarra Regional Food Strategy.* Wollongong, NSW: Ordinary Meeting of Council.

Cherikoff, V. and Isaacs, J. 1989. *The Bush Food Handbook.* Balmain: Ti Tree Press.

DPI (NSW Department of Primary Industries). 2001. *Survey of Recreational Fishing in NSW.* Wollongong, NSW: Department of Primary Industries.

Finch, N., Murray, P., Hoy, J. and Baxter, G. 2014. Expenditure and motivation of Australian recreational hunters. *Wildlife Research,* 41(1), 76–83.

Gasset, J.O.Y. 1995. *Meditations on Hunting.* Belgrade, MT: Wilderness Adventures Press (first published in Spanish 1942).

Green, C. 2009. Managing Laponia: A world heritage site as arena for Sami Ethno-Politics in Sweden. Acta Universitas Upsaliensis. Uppsala: *Uppsala Studies in Cultural Anthropology,* 47.

Gunnarsdotter, Y. 2007. What happens in a Swedish rural community when the local moose hunt meets hunting tourism? in *Tourism and the Consumption of Wildlife: Hunting, Shooting and Sport Fishing,* edited by B. Lovelock. London, England: Routledge, 182–195.

Gurholt, K. 2008. Norwegian *friluftsliv* and ideals of becoming an 'educated man'. *Journal of Adventure Education and Outdoor Learning,* 8(1), 55–70.

Hammarström, T. 2004. *Nordic Giant: The Moose and Its Life.* Bokförlaget: Max Ström.

Hedberg, L. n.d. *Swedish Cooking: Pushing the Limits.* Accessed 19 August 2015. Available at http://www.tryswedish.com/swedish-cooking-pushing-the-limits/.

Isaacs, J. 1987. *Bush Food.* Sydney, Australia: Weldon and Associates.

Jönsson, H. 2013. The road to the new Nordic kitchen – Examples from Sweden, in *The Return of Traditional Food,* edited by P. Lysaught. Lund: Lund University, 53–67.

King, R. 2010. Hunting: A return to nature? in *Hunting: Philosophy for Everyone*, edited by N. Kowalsky. Chichester, England: Wiley-Blackwell, 149–160.

Larsen, H. and Österlund-Pötzsch, S. 2013. Foraging for Nordic wild food: Introducing Nordic island terroir, in *The Return of Traditional Food*, edited by P. Lysaught. Lund: Lund University, 68–78.

Ljung, P., Riley, S., Heberlein, T. and Ericsson, G. 2012. Eat prey love: Game-meat consumption and attitudes to hunting. *Wildlife Society Bulletin*, 36(4), 669–675.

Lopez, B. 2001. The naturalist. *Orion Magazine*. Accessed 19 August 2015. Available at https://orionmagazine.org/article/the-naturalist/.

Louv, R. 2008. *Last Child in the Woods: Saving Our Children from Nature-Deficit Disorder*. London, England: Algonquin Books.

Louv, R. 2011. *The Nature Principle: Human Restoration and the End of Nature-Deficit Disorder*. London, England: Algonquin Books.

Low, T. 1989. *Bush Tucker*. Sydney, Australia: Angus and Roberston.

Łuczaj, Ł., Pieroni, A., Tardío, J., Pardo-de-Santayana, M., Sõukand, R., Svanberg, I., *et al.* 2012. Wild food plant use in 21st century Europe: The disappearance of old traditions and the search for new cuisines involving wild edibles. *Acta Societatis Botanicorum Poloniae*, 81(4), 359–370.

Meyer, C. 2004. *The New Nordic Cuisine Movement*. Accessed 20 August 2015. Available at http://newnordicfood.org/about-nnf-ii/new-nordic-kitchen-manifesto/.

Newling, J. 2011. Dining with strangeness: European foodways on the Eora frontier. *Journal of Australian Colonial History*, 13, 27–48.

Romild, U., Fredman, P. and Wolf-Watz, D. 2011. *Socio-Economic Determinants, Demands and Constraints to Outdoor Recreation Participation in Sweden*. Östersund, Sweden: Friluftsliv i förändring.

Santich, B. 2011. Nineteenth-century experimentation and the role on indigenous foods in Australian food culture. *Australian Humanities Review*, 51(November).

Schulp, C.J., Thuiller, W. and Verburg, P.H. 2014. Wild food in Europe: A synthesis of knowledge and data of terrestrial wild food as an ecosystem service. *Ecological Economics*, 105, 292–305.

Star, A. 2010. Local food: A social movement? *Cultural Studies Critical Methodology*, 10, 479–490.

Stryamets, N., Elbakidze, M., Ceuterick, M., Angelstam, P. and Axelsson, R. 2015. From economic survival to recreation: Contemporary uses of wild food and medicine in rural Sweden, Ukraine and NW Russia. *Journal of Ethnobiology and Ethnomedicine*, 11(1), 53.

Svanberg, I. 2012. The use of wild plants as food in pre-industrial Sweden. *Acta Societatis Botanicorum Poloniae*, 81(4), 317–327.

Tidball, K., Tidball, M. and Curtis, P. 2013. Extending the locavore movement to wild fish and game: Questions and implications. *Natural Science Education*, 42, 185–189.

Vincent, S. 2015. The culling season. *The Monthly*. July 2015.

Waitt, G. 2014. Embodied geographies of kangaroo meat. *Social and Cultural Geography*, 15(4), 406–426.

Wikner, M. 2014. *Jämtland Härjedalen Gastronomy*. Östersund: Jämtland Härjedalen Tourism.

Wikner, M. 2015. *Creative Gastronomy Outdoor Edition*. Östersund: Jämtland Härjedalen Tourism.

8 Managing nature in the home garden

Katarina Saltzman and Carina Sjöholm

> Lawns look nice but require a lot of work. Ours is becoming invaded by dandelions, moss and a small yellow flower that stretches out its tentacles and reproduces kind of like strawberries. It's terrible. I struggle with it everywhere.
>
> (DAG F 1229)

Owners of a private home garden get involved in, and influence, the growth, life and death of other organisms through simple actions like mowing, pruning, trimming and weeding. In the garden, humans inevitably interact with other, non-human actors, such as various plant species. The introductory quote by a Swedish home garden owner reflects the common drama where the gardener and various plants struggle with and against each other. Some home gardeners describe their battle against certain weeds in terms of a continuous war (compare Atchison and Head 2013).

In modern Western societies such as Scandinavia and Australia, the urban home garden is important for hands-on daily interaction with nature's processes. The home garden presents an opportunity to observe the changing of seasons and experience first-hand the growth, fight for survival, maturation and decomposition of other organisms. This chapter discusses everyday interactions with 'nature' in domestic gardens; it is based on personal accounts and has a specific focus on plants. The results and reflections presented in this chapter are outcomes of a cross-disciplinary research project examining the complex interactions between people, plants and other actors of different kinds (such as tools, texts, animals and decorations) in modern Swedish home gardens.[1]

Plants comprise a central component of most home gardens, whether we consider a backyard with a single bush and a lawn or a lush and well-groomed garden with flowerbeds, hedges and vegetable patches. And vegetation, often categorised as weeds, can also take root on surfaces that have been paved to keep the vegetation at bay, for example. Despite their rootedness, plants do move and are moved between different places in the garden, as well as from one garden to another. In the following, we are specifically focussing on the relations between people and plants, especially on the temporal and mobile aspects of plants. We discuss plants as non-human actors (compare Hitchings 2003; Jones and Cloke 2008), which are not only growing but also moving within and between gardens, that is, they refuse to stand still. In many home gardens, there are owners who have stories to tell about how plants have been brought in from a different location, for example, from the garden of an old relative or from a place visited during a holiday. Other plants move by themselves in and between gardens, by spreading seeds, winding roots or rhizomes.

In our project, we have analysed how people interact with their immediate physical surroundings in their home environment. Because the extent to which a home garden can be considered nature is subject to debate among gardeners, humans' relationships with various aspects of nature in the garden give rise to numerous questions. For example, is the home garden considered part of nature or nature part of the garden? Is nature always welcome in the garden, or does it first have to be trimmed, pruned and cleaned up? Who and what is acting in the garden? In order to respond to such questions, we need to attend to issues concerning time, space and social environment. In this chapter, and based on our empirical material, we examine these issues through four larger themes: 'life with a garden', 'cultivated nature?', 'temporality and motion' and 'managing co-species'. However, first we need to put the home garden in context, both empirically and analytically.

The home garden in context

In Sweden, more than half of the total population reside in single-family dwellings (SCB 2008) with access to some form of yard/garden, and many people consider the garden an important aspect of their home environment. The privilege of living in a non-farm single-family dwelling with a garden was once reserved for the more affluent members of the population but spread to other Swedes a little over a century ago (Wilke 2006). Following the end of World War II, subsistence cultivation of fruits and vegetables was gradually replaced with the view of the home garden as a place for leisure and relaxation (Flinck 1994). Subsequently, the market for garden-related items, media and services has grown significantly, and this affects the ways in which people understand and use gardens.

Interestingly, the 'ordinary' private garden has often been taken for granted and rarely problematised in research. In addition, research on cultural processes and components of people's homes has only given limited attention to home gardens (e.g. Miller 2001; Pink 2003; Winther 2006; Shove *et al.* 2007). It is increasingly acknowledged that the home garden is an arena worthy of exploration, however, and it is now a fairly established international research field (e.g. Hitchings 2003; Head and Muir 2006, 2007; Bhatti *et al.* 2009, 2014; Qvenild *et al.* 2014).

In a Swedish context, research on home gardens and their cultural history and significance has so far been surprisingly scarce (Saltzman and Sjöholm 2013, 2014). Swedish garden history research has given considerably more attention to parks and more formal gardens than ordinary home gardens (e.g. Nolin 1999; Ahrland 2005).

Our research project has hence paid particular attention to the home garden, including things people do in the garden without making a big deal of it: for example, the home gardener's daily walk through the backyard to remove overblown flowers, pull out a few weeds or kill a couple of slugs. Routine tasks may include mowing the lawn, pruning trees and watering, but also sitting down comfortably to eat, sunbathe or read. Then there are trips to the nursery to buy seeds, bulbs and summer flowers, but also to the waste recycling centre with twigs, branches and leaves. These are tasks that in a Scandinavian context are associated with changes related to the seasonal cycles.

This chapter is methodologically and theoretically based on ethnological cultural analysis. Ethnology is an interpretive science with similarities to anthropology as well as cultural studies, looking for meaning and understanding in cultural fields. Ethnologists are concerned with exploring everyday life by problematising all sorts of seemingly trivial phenomena, and by studying such mundane activities up-close, cultural analysis can contribute to knowledge about central aspects of our lives. The underlying notion is that people are carriers and active creators of culture (Shove *et al.* 2009; Ehn and Löfgren 2012). By implication, we seek a deep understanding of individual stories and lives. In order to explore individual accounts of what people do in, and say about, their gardens, we have collected material through ethnological questionnaires sent out across Sweden as well as fieldwork including interviews and filmed visits to homes in two residential areas in two communities in southern Sweden.

The use of question lists has a long tradition in Nordic ethnological research (Hagström 2001; Hagström and Marander-Eklund 2009). Their purpose is to gain access to accounts and reflections in relation to everyday experiences (compare Sheridan *et al.* 2000; Bhatti *et al.* 2014). In Sweden, the method has been developed jointly by academic researchers and the folklore archives, which have played an important role in the documentation of past and present everyday culture.[2] As the method encourages informants to write freely using their own words, the responses can vary quite a bit. For example, they may be anywhere from one to dozens of pages in length and can include illustrations and pictures. This chapter is based on responses to two different question lists, one titled *Nature to Me*, which was used in a national documentation project on the roles of nature in everyday life (Midholm and Saltzman 2014), and one titled *The Home Garden*, designed specifically for our project. The latter consisted of questions regarding home gardens in general since not all respondents had their own garden. In total, the two lists have generated about 300 written responses.[3]

We have also conducted ethnographic field studies in two residential areas in southern Sweden. We selected areas where the yards and gardens initially looked the same in order to explore how use of garden space has developed over time. Our case study areas consist of one residential area built in the 1930s and one area developed in the 1960s and 1970s. The fieldwork has consisted of ethnographic observations (Fangen 2005), spontaneous conversation and scheduled semi-structured conversational interviews (Ryen 2004) as well as filmed walk-along interviews (Kusenbach 2003; Pink 2007). We have completed about 40 in-depth interviews. In addition, we have taken pictures, and several informants have shared their own documentation with us. The filming has enabled us to capture some of the small and large changes that take place continuously in the gardens.

In the following sections, we will examine what people show us and say about their gardens.

Life with a garden

The ways plants, objects and environments from the past are approached and handled, in relation to both new trends and the innate dynamics of a home garden, form a recurring theme in the material. Some new homeowners choose to completely remake their yard and garden, while others decide to maintain what is already there. Both approaches can be a significant part of a lifestyle project for the homeowner where the garden becomes an expression of who they are.

We can see movements in space and across boundaries as plants, objects and ideas are moved from one garden to another or from one location in the garden to another. Some of these movements are slow and 'invisible', yet very significant, as when trees grow or new species take hold and spread. In our view, plants can be seen as actors, for example, when they migrate to a garden by themselves, as welcome additions or dreaded weeds. An older woman, now living at a nursing home where she has access to a patio area, talks about a plant that has accompanied her through life:

> When we moved to Silvergatan [in 1945], my uncle Axel brought us a rhubarb root from his allotment. When I moved to Partille [in 1959], my mother gave me a root from the rhubarb. Then when I moved to Gotland [in 1977], I brought the rhubarb roots to my new garden. It became a 10-metre long hedge, and now [2011] I have a small root from the hedge in a pot!
>
> (DAG F 1316)

Many home gardeners tell stories about special plants that came from somewhere else, such as from older family members or from places visited. In this way, the plants become carriers of stories and are assigned special meanings often related to the gardener's personal or family history. One informant with a strong interest in gardening talks about some plants that were moved to her garden. She grew up and lives in southern Sweden and talks about when her parents sold the allotment where she spent all her summers as a child. Her uncle from further north, the only person in her family with a garden, adopted some of the perennials from her parents' allotment. Newlywed and pregnant, she eventually obtained her own garden, to which her parents brought shoots from the perennials that had originally come from the allotment garden. She carefully describes how her parents travelled first to her uncle by public transport and then with a mishmash of trains and buses to her new home in the province of Skåne in southern Sweden with, among other plants, an iris: 'We lived there for about ten years, and the last thing I did before we moved, we were just about to leave ... "Oh I have to bring dad's iris!"' So there she was, in her present, much smaller, garden, showing us what she call 'dad's iris': 'I have it in several places, you're supposed to split it in autumn ... and I've planted it pretty much everywhere' (interview 6).

Maintaining an older garden or special plants that have been moved from older family members' gardens, for example, may represent a very concrete historical link and a manifestation of identity and cultural heritage. Accounts of the origin and development of the garden are common when people talk about or show their gardens and yards, as in the 'dad's iris' example above.

The conservation of an old garden can be a matter of consciously maintaining a piece of green cultural heritage, a practice sometimes held as a responsibility to future generations (Andréasson 2007; Flinck 2013). Aesthetic and economic aspects may also be relevant, as an old-fashioned garden may be an asset when selling a house and a garden. However, the owner of a garden can never be certain a new owner will appreciate what he or she has accomplished. A garden does not exist only at a physical level but also as ideas in the owner's mind, and it is obvious that different garden ideals operate in parallel. The notion that the garden is a perhaps not perfect, but at least pleasant, reverberation of a lost paradise implies still that dreams often trump the actual outcome (Gunnarsson 1992). Thus, ideals and practice merge in the garden, and many garden projects may contribute most to the owner's enjoyment while still only imagined.

Cultivated nature?

Many researchers have pointed to the problem of considering nature and culture as two completely separate spheres, as it leads us to think of objects, plants, people, ideas and so on as belonging to one or the other of these categories. This limits our ability to see and understand boundary-crossing relationships and processes, according to researchers such as David Harvey (1996), Sarah Whatmore (2002), Bruno Latour (2004) and Donna Haraway (2008). The word 'nature', however, is firmly rooted in both the knowledge tradition of Western society and in more trivial understandings of the world around us. Our research confirms that the concept pair nature/culture is a strong figure of thought and an important part of many people's understanding of their everyday reality, not least in the garden.

Home gardeners display great diversity in their views of the extent to which the natural elements in their gardens can and should be cultivated and controlled (Pollan 1991). Some gardens can be seen as a manifestation of human control of nature. Other garden owners prefer an environment that resembles nature as much as possible. According to one informant, a retired farmer who now devotes much time to her garden:

> The ultimate goal is to have a natural-looking garden, one that looks like nature itself. There are no straight lines in that type of garden. Things are planted and allowed to grow wherever there is a piece of bare soil. It's not easy, let me tell you, it's a lot harder than making straight lines, and I think I have created a wonderful piece of 'nature' here. You need to keep on top of the weeds, of course, and the lawn has to be mowed.

(LUF M 25977)

Having the garden resemble nature is a strong and widespread ideal among gardeners (Kingsbury 2006) and gives rise to many questions among home gardeners. Issues gardeners struggle with include whether or not to apply fertilizers, prune and mow and whether to introduce new plants or only work with what happens to shoot out of the ground. The question of how and how much the natural environment should be controlled yields different answers depending on who and when you ask.

Managing a home garden is also a matter of learning to cooperate with nature. In recent years, the home garden has increasingly been recognised as an important environment for biological diversity, where a lot can be done to benefit wild plants and animals (e.g. Gaston *et al.* 2007; van Heezik *et al.* 2012). Although some informants expressed firmly that a garden is not nature, the distinction between what is perceived as nature and culture is rather fuzzy. Several informants wrestle in their responses with questions of an almost philosophical kind about where nature begins and ends (Figure 8.1). Hence, it can be difficult to determine the boundaries of what a garden is and what it is not. After all, what is it that makes a piece of land a garden?

Figure 8.1 The owners of this house cherish, cultivate and collect 'wild' plants such as raspberries on the public parkland hillside adjacent to their garden.

On the outskirts of one of the residential areas, we interviewed a couple with small children whose garden neighbours a hill that is publicly owned parkland. This interview illustrated a variety of permeable boundaries within a single garden, including wild/cultivated, order/disorder as well as private/public. Here, we became aware that something that may appear as disorder and neglect on the surface can in fact be a result of the gardener's specific ambitions and understandings of order. The couple told us that they had 'sown wild plants' and both have an interest in what they call medicinal plants. 'I want to have plants that are useful, that are good, that we can take advantage of', said the woman in the family. She added that an inventory of the flora in the parkland had identified some protected plant species close to her garden, implying that these might have spread from their garden (compare Head and Muir 2006). This family actually regarded a significant part of the hillside behind their backyard as a part of their garden. And so, it seems, had previous owners, ever since the house was built in the 1930s. In the interview the current owner, who also grew up in this house and garden, reflected on this: 'It feels like my piece of land. I was a bit surprised when I learned that it was not part of my property, because I have always regarded it as my land' (interview 5).

It can be difficult to draw a line between what constitutes a garden and what does not, and a range of answers to what makes a garden is possible. Many garden owners see boulders, large trees, a creek or a natural pond as attractive features of a garden and are willing to invest considerable resources in incorporating such elements in environments where they are lacking (Londos 2004). Others take the natural features of the garden for granted. Other responses, such as this one from the west coast of Sweden, indicate a clear distinction between more 'natural' and more 'cultivated' parts of the garden.

> A few words about the yard around my house. I have a mixed yard measuring about 2 000 square metres. I have a big lawn and several flowerbeds, but leave parts of the yard as-is. Looking at what nature has to offer in these parts, there are alder, elm, ash and birch, as well as an enormous poison ivy covering the bottom 10 metres of the trunk of an ash tree. ... I enjoy the variation between natural and cultivated elements, in particular since it is my home.
>
> (DAG F 1210)

Many respondents talk about experiences related to the natural environment they have acquired in the garden. These accounts often concern the hands-on confrontation with all the things that grow in their gardens. They talk about enjoyment and recreation, but also about maintenance and weeding. Our informants' examples help shed light on the ambiguous distinction between desirable and undesirable aspects of the garden, including what happens when plants grow too numerous or too large. One interviewee describes the transformation of her garden by explaining how one task – mowing the lawn – after a couple of years was replaced with another – weeding – when she redesigned her small yard by replacing the lawn with several rock gardens. The weeding does not bother her much since she remembers the mowing as a major nuisance. She talks about strategies to minimise weeding, though, as she allows many plants to spread and cover the soil:

> A lawn requires a lot more work than you think. If you want a nice one, that is. And I did, of course. Yeah. So I mowed it, and raked it, and mowed it again and I also brushed it. ... So it wasn't easy, but this is not too bad. I pick weeds. ... And then, all the plants cover the ground so the weeds don't stand a chance.
>
> (Interview 1)

The lawn is often discussed as a symbol of the cultivated aspects of the garden. Whether a lawn is a low- or a high-maintenance project is subject to debate. In the end, it seems to be a matter of ambition. Lawns seem to be assigned special importance in many parts of the world, and this has been discussed, not least by Paul Robbins (2007), who has examined the lawn from ecological, economic and social perspectives, and Russel Hitchings' (2006) study of 'retreating lawns' in the UK. Without being directly asked about it, many informants talk about the maintenance of the lawn but also about a weed-free lawn as an ideal. Few informants seem to be content with a lawn 'as long as it's green'. A lawn full of weeds and moss is frequently referred to as a problem, yet not everybody agrees: 'I guess we're a bit Japanese in that way, we actually like the moss', says one informant (interview 2). Some have sacrificed flowerbeds and vegetable patches for a bigger lawn, often with reference to their children's needs and activities: 'We wanted a larger lawn for the children. We used to have just two small patches of grass. We threw away all the berry bushes and removed a path and a vegetable patch' (LUF M 26211). This is one example of how the home garden evolves through adaptation to the changing phases of family life, which leads us to the issues of time and change in the garden.

Temporality and motion in the garden

> I can enjoy nature in my small and not always well-kept garden every single day. Sometimes I can see clearly how things grow and the garden changes from day to day, but other times I just don't notice what happens for several weeks. When you come back after a week away, it's easy to see that a lot can happen in a short time. The garden also shows clearly that nature lives its own life.
>
> (LUF M 25943)

Gardens involve many movements and motions; changes connected to the cycles of the year, of day and night, life cycles of individuals, intermingled and combined with decisions and actions of human and non-human actors. Within the field of 'more than human geographies', Lesley Head and Jennifer Atchison (2009) have recognised a growing interest in the hybridity (compare Whatmore 2002) and fluidity of human–plant relations. Similarly, within the field of anthropology, Tim Ingold and Gisli Pálsson (2013) argue for an understanding of all organisms – human and non-human – not as bounded entities but as biosocial becomings that can be understood as ensembles of relations in flows of materials. All life, Ingold argues, has in common that it is simultaneously social and biological.[4] And living organisms are better understood as trajectories of movement and growth (that is, becomings) than as discrete, preformed entities (that is, beings). These perspectives direct our attention away from individuals and objects and invite us to focus instead on flows and motions or, as Ingold phrases it, 'to think of humans and indeed of creatures of all other kinds, in terms not of what they are, but of what they do' (2013, p. 8).

The garden contains clearly cyclical changes, leading to various motions and transformations. Seeds are sown and plants planted; they grow and flourish, wilt and may face their last days in the compost bin. Some of these movements are fast and lead to considerable change, while others are slower and pass unnoticed. Our material gives many indications that the home garden can be a place for everyday rituals and routines, and in connection with these accounts, many informants reflect on various aspects of time and change. As mentioned, some garden owners take a daily walk through the garden, at least in the warmer half of the year, either as a morning routine or to wind down after a day at work. They often describe how this enables them to follow the yearly cycle of growth, blooming and wilting, day by day and week by week (Figure 8.2). Swedish garden owners see the changing of seasons as a basic component of their experience of the garden, and the yearly cycle is often emphasised in descriptions of seasonal tasks in the garden. Some

Figure 8.2 Eagerly waiting for her wood anemones (*Anemone nemorosa*) to flower; this gardener almost tries to open their buds with her hands.

informants describe how they feed birds and squirrels every winter, while others make it a tradition to gather Christmas decorations from the garden. However, most people seem to take a break from their gardens in the wintertime and may instead spend their time making plans for the next gardening season. When the spring finally arrives, the results of the previous year's efforts may show:

> Last autumn I planted flower bulbs for the first time ever, and have enjoyed watching all the grape hyacinths and tulips pop out of the ground. I also enjoy picking weeds. It's a never-ending job and somehow seems meaningless, but I love it! Pulling them out of the soil, making it look nice, it smells fresh somehow.
>
> (LUF M 25935)

This informant stresses how she enjoys the 'meaninglessness' of weeding. In fact, the seemingly meaningless tasks in a garden appear to offer many garden owners a special kind of meaning. They pull out weeds, the weeds grow back, they pull them out again, the cycle never ends, and that is precisely what makes it meaningful. The repetitive tasks in a garden become a way to connect with the cycles and processes of nature. The garden and its many organisms are in a constant process of change, resembling the contexts of life, growth, ageing and death of which we are all part.

Many informants laugh in recognition when asked about weeds. 'It's a never-ending job. So I don't care anymore, I just mow the lawn before they seed. ... But you simply can't get rid of them' (interview 3). Weeding is something garden owners just do and are expected to do. Our material includes many descriptions of different types of weeds as well as strategies to minimise their presence in the garden, a topic we will return to in the next section. Some plants grow too big and the magic line between too big and too little becomes a problem to many garden owners. Some informants also talk about their mistakes. 'We have a problem with this walnut tree. We really want to keep it, it keeps growing, it's too big. We're not sure what to do. It's very productive, you know' (interview 4). Trees and bushes end up much larger than planned and a hedge harder to trim than expected. A hedge might even turn into an unplanned impermeable green wall.

Our material also includes interesting descriptions of how new plants necessitate removal of old ones and how this in turn affects other plants. Some gain 'new vitality', as described by one informant (LUF M 26236). The same informant's awareness of the different roles of plants becomes evident when she concludes that some plants 'take a lot of space and really don't look very good, but are indispensable on hot summer days when the awning and magnolia don't cut it on the patio' (LUF M 26236).

Sometimes the changes in a garden are linked to other events in life, as it is not uncommon to plant a tree to commemorate the birth of a child or the passing of a family member. Plants hence carry memories; they may have been received as a gift, shared with friends or acquired in a special place. For many garden owners, the garden may be considered a social project in which an exchange system is of central importance (compare, for example, Belk 2013). The practice of trading plants with other people is not new, but the extent and opportunities have increased and plant exchange is not limited to specialists. For example, today home gardeners can easily obtain cuttings, seeds and split perennials either online or at special 'plant flea markets'.

Managing co-species

The garden is clearly an arena where humans and other organisms interact in reciprocal relationships, involving the gardener and other people, as well as plants and non-human animals. It is also true that the garden is created through both refinement of and a battle against nature. As demonstrated, 'nature' can refer to a certain part of a garden, an aesthetic or environmental ideal, the wild and untamed – pests and weeds threatening other plants – but at the same time also a notion of a greater context of which the garden is part. One of many issues garden owners face concerns how much the garden should be controlled and arranged (Figure 8.3). Sometimes it is obvious which plants should be kept or eliminated, but this decision may also be a matter of personal preference: 'I let the plants grow wherever they want and don't care if they end up in straight lines or not. Some pop up all by themselves – I've never planted tansy and foxglove, but they're there anyway' (LUF M 26033).

Many people associate home gardening with dreams of the simple, good life 'close to nature'. In practice, however, nature tends to offer plenty of resistance, and the work in the garden can often be described as a fierce battle. Hence, from this perspective it is of particular interest to reflect on issues relating not only to 'weeds' but also to 'alien species':

> Garden work is not my thing. Maybe because of the Spanish slug and ground elder. I used to plant flower seeds, vegetables etc., but when the pests get all of it, you lose interest. Every single evening after it rains in the summer, I go out in the garden with a hand trowel and an empty milk carton. I chop them (the slugs) in half and put them in the carton. Sometimes I count them; one evening I got 150. I don't think it makes much of a difference, though, since I don't think my neighbours are doing anything about them.
>
> (LUF M 26222)

Whether an organism in the garden is considered an asset or a threat is determined through cultural and social processes. This becomes particularly noticeable in cases where neighbours have different attitudes to, for example, ground elder (*Aegopodium podagraria*) or the Spanish slug (*Arion vulgaris*), infamous intruders in Swedish gardens that can spread rapidly across property lines. It is also important for a garden owner to learn which plants may become too comfortable in the garden. A woman who has lived in her house and garden for 40 years mentions cicely (*Myrrhis odorata*) and oregano (*Origanum vulgare*): 'You need to be strict with the cicely or else it takes over the whole garden. So does the oregano (wild marjoram), but the butterflies love it' (LUF M 26033). When these plants act and spread in her garden, she is ready to interact with them and interfere with their motions.

Figure 8.3 For many people, an important aspect of gardening is order and control.

Many informants talk about plants popping up here and there, often in large quantities, and that they do not know what it is, where it comes from and how to handle it: 'It pops up by itself, a lot of it, I don't know what it is, ... it just keeps coming, you have to keep ripping it up' (interview 4). Unexpected and uncontrolled spreading is usually described as a problem but sometimes also as something intriguing of which the garden owner may even take advantage. Ground elder is edible, and so are, for example, the nettles (*Urtica dioica*) that thrive in some gardens. For some, the presence of these plants would be a problem, while for others it is not. For example, one gardener who grows a lot of vegetables mentions how the nettles become a seasonal delicacy (interview 2). Another who claims that she is 'definitely not a meticulous gardener' appreciates the hollyhocks' (*Alcea rosea*) seed setting: 'A nice colony of hollyhocks has grown up along the external wall (a hopeless location for most plants!). They take care of their own reproduction, too. You just have to pull up the ones that come up in the wrong places' (LUF M 26236). Some informants point to plants they like but do not have a clue where they came from. A couple of devoted gardeners (interview 2; LUF M 26194) talk about how they like to dig up or pick seeds from plants that they want and that grow outside other people's yards (compare Phillips 2013). Others mention plants in their own gardens that tend to spread. One example of a plant that can be considered both a weed and a decorative addition is the creeping bellflower (*Campanula rapunculoides*), a bluebell that spreads easily: 'I have a hard time deciding whether it's a beautiful flower I want to keep or an annoying weed that's impossible to get rid of. I think I'll at least try to limit the spreading of it' (LUF M 26236).

Some plants grow out of control and even become invasive if allowed to grow freely in an area they like. While in Australia the spreading of invasive, non-native plant species is a widely recognised problem (Atchison and Head 2013; Head 2014), this is not as much acknowledged in the Scandinavian countries, with the possible exception of Norway (see Setten, this volume) where a 'black list' has existed since 2010 (Qvenild 2013; Qvenild *et al.* 2014). In both Australia and Europe, many species that have turned out to be invasive have been spread via cultivation in home gardens. The European Union has started working on a list of invasive alien species since a new regulation (1143/2014 on invasive alien species) came into force in 2015. The European network for invasive alien species, NOBANIS, lists a total number of 387 invasive alien species in Sweden as of May 2016 (www.nobanis.org). The Swedish Environmental Protection Agency has been involved in the listing, inspired by the Norwegian 'black list'. Among the species suggested for the list, we find many that are mentioned and grown by our informants, such as lupine (*Lupinus polyphyllus*), loosestrife (*Lysimachia punctata*), lilac (*Syringa vulgaris*), and elderberry (*Sambucus nigra*). Many people pick the elderflowers to prepare a popular sweet drink. A retired man with a large rural garden in southwest Sweden considers the spread of elderberry in the area a sign of a more general change in the landscape:

> I consider the elder bushes a weed. They are everywhere. I have shaped a few of them to nice-looking trees. When we moved here in the mid-70s, there was no elderberry anywhere. We had to go to Hjärnarp to pick flowers for the drink. The spread of it, and of cow parsley and soft rush, is the change in the vegetation I've been most concerned about in our 36 years on Hallandsåsen.
>
> (LUF M 26203)

This brings our attention to a dimension of change not many informants addressed: general changes in the flora and the landscape which can be both local and global.

Let us return to the couple who had incorporated part of the neighbouring parkland into their garden and sown 'wild' plants. Even for the owner of this garden, who otherwise cherishes everything she understands as 'natural', all plants are not welcome in the garden:

> I have been the first to combat ground elder in this area. I have removed all *their* ground elder, *their* ground elder and *theirs*. I do not like it, because it is taking over so much. I have nothing against the dandelions, they are healthy and good. And actually, so is ground elder, but it takes over too much.
>
> (Interview 5)

In his book *Second Nature* (1991), Michael Pollan discusses the modern Westerner's relationship with nature based on his own gardening experiences. It is both more difficult and more important, says Pollan, to find an ethical approach to nature when you literally are participating in it, as in a garden, than when you observe an unadulterated wilderness from a distance. In a garden, it is hardly reasonable to claim that nature always fares better if humans abstain from tinkering with it. 'The gardener tends not to be romantic about nature. What could be more natural than the storms and draughts and plagues that ruin his garden?' writes Pollan and concludes that the home gardeners therefore also feel they have the right to quarrel with nature (1991, p. 192ff).

Several informants talk about how they welcome the wild and untamed nature with open arms, including various animals, in their gardens. One informant says she enjoys looking out of her sitting room window at the deer that come to visit in winter evenings to eat from the bird feeder, but that she shoos the same deer away later in spring when they want a bite of her tulips. Thus, what is welcome can vary over time and depending on the situation, which is another example of how diverse and contradictory our relationship with what we call nature can be. It is not uncommon that gardeners see their plans and dreams shattered by some four-legged visitors:

> The other day when I was looking up something in our old gardening book, I found my husband's crossed-out notes about what was supposed to become our fruit garden. Apple and plum trees were purchased in 1987 and 1988: Reine Claude, Victoria, Gyllenkroks, Transparent Blanche, Silva, Oranie, Lobo, Cox's Pomona and Åkerö. An elk put an end to our plans not once but twice.
>
> (DAG F 1229)

Some other informants talk passionately about the birds they feed: 'They return what they receive in the winter, [by] eating the pests in the summer' (interview 7). This view, that birds and humans help each other in the garden, reflects an attitude we have come across repeatedly. In the everyday interaction that takes place in the garden with other organisms, both animal and plant co-species (compare Haraway 2008), many people seem to experience a sense of affinity and mutual purpose that extends across species lines.

The garden is often described as the quintessence of our encounter with and cultivation of nature. In contemporary home gardens, it is obvious that plants and animals continuously trespass on attempts to establish boundaries between nature and culture. In this section we have seen how the approaches of garden owners to organisms in their gardens unveil boundaries and categorisations that are important, yet often unreflected, in everyday life in the garden.

Concluding reflections

So, what is there to learn from studies – at this very local landscape level – of interactions with mobile plants in home gardens? In this chapter we have examined some interrelations and biosocial processes through which nature is interpreted, delineated and formed in the context of contemporary home gardens. Based on our informants' accounts, we can conclude that the home garden is a very important place for everyday interactions with, and negotiations around, nature. To live with a garden is to influence and be influenced by an environment; to form it and be formed by it. Raking leaves, weeding and pruning trees can be seen as a concrete way of controlling 'nature' by creating and maintaining some degree of order and structure. What people have learned to define as nature is continuously managed and shaped in the garden – to a greater or lesser extent – yet it is never fully subject to human control. A garden that has grown out of control can be considered either a failure or an ideal, depending on the perspective taken.

Our ethnographic methods have enabled a close-up examination of the understandings and management of nature by individual garden-owners in their own gardens. Their stories and actions show that categories such as nature and culture are in this context highly relative. They remind us also that the diversity of strategies, interpretations and negotiations that people make in relation to a variety of non-human actors always has to be taken into account.

We have paid particular attention to the informants' perspectives on plants and their movements, and we have observed the agency of spreading plants and their trajectories of motion and growth, as they move within and between gardens. We have seen that some plants in some gardens are regarded as useful and pretty, while in other gardens they are despised as weeds. By looking closely at what our informants have to say about plants and other components of the dynamic microcosm of the garden, we have seen how people relate and respond to temporality and complex processes of change. The life cycles of both people and plants create rhythms in the garden, related to diurnal, seasonal and lifetime changes. With new phases in the garden-owner's life come new priorities that are likely to affect their relationship to, and management of, the garden and its plants. Similarly, plants do develop and affect their human and other co-beings in new ways over time. In the garden, biosocial humans and biosocial plants constantly interact with each other, as well as with other actors and becomings.

The complex, dynamic, multi-species conglomerate of the garden can easily be interpreted in terms of hybridity, fluidity and boundary-crossing biosocial becomings. This kind of terminology certainly provides a more nuanced analytical framework than simplifying binaries such as nature/culture. At the same time, nature and culture is clearly a viable figure of thought in everyday practice, strongly maintained, for example, in the home garden. Hence, we need to remember that these binary concepts have a strong hold on understandings not only of large-scale and philosophical issues but also of everyday matters in our closest vicinity.

Notes

1 The project *Work and Tools in the Garden of Dreams and Realization*, funded by the Swedish Research Council, was carried out in 2012–2015. As such, it reflected primarily the situation in the 2010s, even if many of our informants talked about their lives with their gardens in a longer perspective. Allan Gunnarsson, landscape architect at the Swedish University of Agricultural Sciences, also participated in the project. We thank Johan Hultman for his helpful comments on this text. Parts of the material

presented in this chapter have previously been published in Swedish (Saltzman and Sjöholm 2013, 2014).

2 The Swedish folklore archives utilise a fixed group of informants who on a voluntary basis answer thematic questions in a wide range of areas related to everyday life and experiences.

3 The chapter includes quotes from the question list material with references to the archives where the responses are kept: Department of Dialectology, Onomastics and Folklore Research in Gothenburg (DAG) and the Folklore Archives in Lund (LUF).

4 Ingold even argues that the 'domains of the social and the biological are one and the same' (2013, p. 9).

References

Ahrland, Å. 2005. *Den osynliga handen: Trädgårdsmästaren i 1700-talets Sverige*. Diss. Stockholm: Carlsson.

Andréasson, A. 2007. *Trädgårdshistoria för inventerare*. Alnarp, Sweden: Centrum för Biologisk Mångfald.

Atchison, J. and Head, L. 2013. Eradicating bodies in invasive plant management. *Environment and Planning D: Society and Space*, 31, 951–968.

Belk, R. 2013. You are what you can access: Sharing and collaborative consumption online. *Journal of Business Research*, 67(8), 1595–1600.

Bhatti, M., Church, A., Claremont, A. and Stenner, P. 2009. "I love being in the garden": Enchanting encounters in everyday life. *Social & Cultural Geography*, 10(1), 61–76.

Bhatti, M., Church, A. and Claremont, A. 2014. Peaceful, pleasant and private: The British domestic garden as an ordinary landscape. *Landscape Research*, 39(1), 40–52.

Ehn, B. and Löfgren, O. 2012. *Kulturanalytiska verktyg*. Malmö, Sweden: Gleerup.

Fangen, K. 2005. *Deltagande observation*. Malmö, Sweden: Liber Ekonomi.

Flinck, M. 1994. *Tusen år i trädgården: Från Sörmländska herrgårdar och bakgårdar*. Stockholm, Sweden: Tidens Förlag/Torekällbergets Museum.

Flinck, M. 2013. *Historiska trädgårdar: Att bevara ett föränderligt kulturarv*. Stockholm, Sweden: Carlssons.

Gaston, K.J., Fuller, R.A., Loram, A., MacDonald, C., Power, S. and Dempsey, N. 2007. Urban domestic gardens (XI): Variation in urban wildlife gardening in the United Kingdom. *Biodiversity and Conservation*, 16(11), 3227–3238.

Gunnarsson, A. 1992. *Fruktträden och paradiset*. Diss. Alnarp: Sveriges Lantbruks universitet.

Hagström, C. 2001. Local informants and collaborators: Examples from the Folklife Archive in Lund, in *Input and Output: The Process of Fieldwork, Archiving and Research in Folklore*, edited by U. Wolf-Knuts. Turku, Finland: Nordic Network of Folklore, 25–38.

Hagström, C. and Marander-Eklund, L. (eds). 2009. *Frågelistan som källa och metod*. Lund, Sweden: Studentlitteratur.

Haraway, D.J. 2008. *When Species Meet*. Minneapolis, MN: University of Minnesota Press.

Harvey, D. 1996. *Justice, Nature and the Geography of Difference*. Oxford, England: Blackwell.

Head, L. 2014. Living in a weedy future: Insights from the garden. *Bulletin för Trädgårdshistorisk Forskning*, 27, 10–13.

Head, L. and Atchison, J. 2009. Cultural Ecology: Emergent human-plant geographies. *Progress in Human Geography*, 33(2), 236–245.

Head, L. and Muir, P. 2006. Suburban life and the boundaries of nature: Resilience and rupture in Australian backyard gardens. *Transactions of the Institute of British Geographers*, 31(4), 505–524.

Head, L. and Muir, P. 2007. *Backyard. Nature and Culture in Suburban Australia*. Wollongong, Australia: University of Wollongong Press.

Hitchings, R. 2003. People, plants and performance: On actor network theory and the material pleasures of the private garden. *Social & Cultural Geography*, 4(1), 99–114.

Hitchings, R. 2006. Expertise and inability: Cultured materials and the reason for some retreating lawns in London. *Journal of Material Culture*, 11(3), 364–381.

Ingold, T. 2013. Prospect, in *Biosocial Becomings: Integrating Social and Biological Anthropology*, edited by T. Ingold and G. Pálsson. Cambridge, UK: Cambridge University Press, 1–21.

Ingold, T. and Pálsson, G. (eds). 2013. *Biosocial Becomings: Integrating Social and Biological Anthropology*. Cambridge, UK: Cambridge University Press.

Jones, O. and Cloke, P. 2008. Non-human agencies: Trees in place and time, in *Material Agency. Towards a Non-Anthropocentric Approach*, edited by L. Malafouris and C. Knappett. Boston, MA: Springer, 79–96.

Kingsbury, N. 2005. As the Garden so the Earth: The politics of the 'Natural Garden', in *Vista: The Culture and Politics of Gardens*, edited by T. Richardson and N. Kingsbury. London, UK: Frances Lincoln.

Kusenbach, M. 2003. Street phenomenology: The go-along as ethnographic research tool. *Ethnography*, 4(3), 449–479.

Latour, B. 2004. *Politics of Nature. How to Bring the Sciences into Democracy?* Cambridge, MA: Harvard University Press.

Londos, E. 2004. Fem trädgårdar som samhällsspegel, in *I skuggan av Gnosjöandan*, edited by E. Londos. Jönköping, Sweden: Jönköpings Läns Museum, 169–178.

Midholm, L. and Saltzman, K. (eds). 2014. *Naturen för mig: Nutida röster och kulturella perspektiv*. Göteborg, Sweden: Institutet för Språk och Folkminnen.

Miller, D. (ed.). 2001. *Home Possessions: Material Culture behind Closed Doors*. Oxford, England: Berg.

Nolin, C. 1999. *Till stadsbornas nytta och förlustande: Den offentliga parken i Sverige under 1800-talet*. Diss. Stockholm: Univ.

Phillips, C. 2013. *Saving More than Seeds*. Farnham, England: Ashgate.

Pink, S. 2003. *Home Truths: Gender, Domestic Objects and Everyday Life*. Oxford, England: Berg.

Pink, S. 2007. *Doing Visual Ethnography. Images, Media and Representation in Research*. 2nd edition. London, England: Sage.

Pollan, M. 1991. *Second Nature: A Gardeners Education*. New York, NY: Grove Press.

Qvenild, M. 2013. *Wanted and Unwanted Nature: Invasive Plants and the Alien-Native Dichotomy*. Diss. Trondheim: Norwegian University of Science and Technology.

Qvenild, M., Setten, G. and Skår, M. 2014. Politicising plants: Dwelling and alien invasive species in domestic gardens. *Norsk Geografisk Tidsskrift-Norwegian/Journal of Geography*, 68(1), 22–33.

Robbins, P. 2007. *Lawn People: How Grasses, Weeds, and Chemicals Make Us Who We Are*. Philadelphia, PA: Temple University Press.

Ryen, A. 2004. *Kvalitativ Intervju – från Vetenskapsteori till Fältstudier*. Malmö, Sweden: Liber ekonomi.

Saltzman, K. and Sjöholm, C. 2013. Gräva upp och klippa ner – liv och rörelse i Villaträdgården, in *Proceedings from ACSIS conference On the Move, Norrköping 11–13 Juni 2013.* Norrköping, Sweden: Linköpings Universitet.

Saltzman, K. and Sjöholm, C. 2014. Trädgård – kultiverad natur? in *Naturen för mig. Nutida röster och kulturella perspektiv,* edited by L. Midholm and K. Saltzman. Göteborg, Sweden: Institutet för språk och folkminnen, 239–251.

SCB (Statistiska Centralbyrån). 2008. *Bostads- och Byggnadsstatistisk Årsbok 2008.* Örebro, Sweden: Statistiska Centralbyrån, Enheten för Byggande, Bostads- och Fastighetsstatistik.

Sheridan, D., Street, B.V. and Bloome, D. 2000. *Writing Ourselves: Mass-Observation and Literacy Practices.* Cresskill, NJ: Hampton Press.

Shove, E., Trentmann, F. and Wilk, R.R. (eds). 2009. *Time, Consumption and Everyday Life: Practice, Materiality and Culture.* Oxford, England: Berg.

Shove, E., Watson, M., Hand, M. and Ingram, J. 2007. *The Design of Everyday Life.* Oxford, England: Berg.

van Heezik, Y.M., Dickinson, K.J.M. and Freeman, C. 2012. Closing the gap: Communicating to change gardening practices in support of native biodiversity in urban private gardens. *Ecology and Society,* 17(1), 34.

Whatmore, S. 2002. *Hybrid Geographies: Natures, Cultures, Spaces.* London, England: Sage.

Wilke, Å. 2006. *Villaträdgårdens Historia: Ett 150-Årigt Perspektiv.* Stockholm, Sweden: Prisma.

Winther, I.W. 2006. *Hjemlighed: Kulturfænomenologiske Studier.* København, Denmark: Danmarks Pædagogiske Universitets Forlag.

Electronic sources

www.nobanis.org

Archival sources

Responses to questionnaires:

Department of Dialectology, Onomastics and Folklore Research in Gothenburg (DAG): Responses to the question lists DAG 20 Nature to Me (Naturen för mig) and DAG 22 The Home Garden (Trädgården). Archival series F.

The Folklore Archives in Lund (LUF): Responses to the question lists LUF 230 Nature to Me (Naturen för mig) and LUF 233 The Home Garden (Trädgården). Archival series M.

Interviews:

Recordings and transcriptions are kept by the authors.

Part III

Indigenous challenges to environmental imaginaries

9   Indigenous land claims and multiple landscapes

Postcolonial openings in Finnmark, Norway

Gro B. Ween and Marianne E. Lien

Introduction

Norway and Australia are nation-states where current land practices and rights claims speak to contested natures and notions of belonging. This is particularly salient in relation to recognition by these nations of their respective Indigenous populations. Since the 1970s, both countries have seen legal and political processes, rectifying practices in which postcolonial frameworks have been sought, negotiated and articulated against a background of a gradual recognition of colonial pasts. Such postcolonial practices serve to imagine, re-imagine and align pasts and futures in relation to Indigenous peoples and the landscapes and places they inhabit.

This chapter explores an ongoing process of establishing local and Indigenous user rights in Finnmark, northern Norway. Our concern is the extent to which this ongoing process allows for what we tentatively refer to as 'otherness within', that is, the nation's acknowledgement of multiple natures and/or of Indigenous and local nature practices that are radically different from hegemonic and legal notions of property. In our analysis, we explicitly engage Australian legal processes as a comparative figure. Our aim is not to provide an exhaustive account of Australian native title processes but rather to use this as a lens for examining the Norwegian Indigenous legal process. We have observed that while legal practices framing landscapes and indigenous rights are hardly straightforward anywhere, the Australian native title framework appears, at certain moments at least, to encompass the possibility of multiple natures, as it has incorporated attempts to engage an Indigenous other with radically divergent practices of nature and time (see, e.g., Verran 1998; see also Stengers 2005). Our concern then is the extent to which postcolonial openings, provided by an acknowledgement of multiplicity, are incorporated in the ongoing Sami rights process in Norway.

In this chapter, we analyse the legal and bureaucratic processes through which different legal trajectories emerge and discuss their respective procedures, as well as the extent to which each allows for the articulations of multiple natures. We are inspired by recent dialogues concerning attempts to make space for what de la Cadena refers to as a world where many worlds fit (2015). Emerging from critique of the distinction between nature and culture, and against what is often referred to as the 'one-world world' (Law and Lin 2011), this is an ontological move that may encompass – but does not necessarily require – radical alterity.[1] Rather it reflects a concern with what might be referred to as 'the uncommons' (de la Cadena 2015).

The term *uncommons* is an attempt to articulate the possibilities for alliances (between local guardians of land and environmental activists, for example) based on uncommonalities, that is, alliances or negotiations that do not undo fundamental differences but may incorporate 'the parties' constitutive divergence – even if this opens up discussion of the partition of the sensible and introduces the possibility of ontological disagreement into the alliance.' De la Cadena continues:

> this alliance would also house hope for a commons that does not require the division between universal nature and diversified humans: rather, a commons constantly emerging from the uncommons as grounds for political negotiation of what the interest in common—and thus the commons—would be.
>
> (2015, n.p.)

Hence we ask: what would it take to articulate Indigenous notions of landscape belonging and land rights in a way that allows such uncommons to co-exist? To what extent is there currently a space for articulating divergences that undo a singular nature, or what Anna Tsing has referred to as processes of worlding – that is, 'the always experimental, partial, and often quite wrong, attribution of worldlike characteristics to scenes of social encounter' (Tsing 2010, p. 54).

In both Norway and Australia, the nation-states have historically lacked recognition of Indigenous ownership of land. The colonial nature of these historical legal doctrines became publicly visible at more or less the same time in the two countries, times when politics in both places were characterised by strong egalitarian ideals and prevalent ideals of fairness. In Australia, federal land rights for Indigenous peoples became known as Native Title in 1993. In Norway, the official legal process was lengthier and was only formalised in 2005, with the Finnmark Act and the subsequent establishment of the Finnmark Estate (*Finnmarkseiendommen*). In both nations, the land made available for Indigenous title was common land ('crown land' in Australian), land not subject to private ownership.

The Finnmark Act differs in some important ways from the Australian Native Title Act. First, the Finnmark Estate (FEFO) represented a national legal process, transferring ownership of our northernmost county to an Estate. This legal process, however, was of little consequence to Sami people outside this county. Second and more important to our argument, while the handover of land from the state to the Finnmark Estate acknowledged an Indigenous Sami population and its rights to land, the Finnmark Estate is not a Sami Estate. The Estate was explicitly handed over to the entire population of Finnmark, both Sami and non-Sami. The new Estate is also managed by a board of six directors, only three of whom are appointed by the Sami Parliament of Norway, while the other three are named by the Finnmark County Council. As Sami have the right to vote both in the Sami Parliamentary elections and the county council elections, all directors could hypothetically be Sami, but not necessarily. The population in Finnmark is Sami, Kvæn[2] and Norwegian. Stigma experienced after years of state assimilation policies have a long-term effect. This, and the fluid and negotiable nature of ethnicity and kinship in this region, has manifested in families over several generations and led to situations in which different members ascribe to all three kinds of ethnic identification (Kramvig 2005). Therefore, to determine 'who is who' in Finnmark is by no means straightforward. Also, Norwegian political parties as well as Sami political parties are represented at the county council, and ethnic identity and political affiliation do not necessarily coincide. As we have noted elsewhere (Ween and Lien 2012), the Finnmark Act recognises the rights of the entire population of Finnmark, and a multiethnic population is thus institutionally incorporated in the management structure.

This brief comparative background should be kept in mind as we turn to a close examination of the actual local work of recognising local user rights, or potentially property rights, the ongoing legal processes of the Finnmark Commission. In the Norwegian case, this work was not initiated before 2008, several years after the land transfer. In this chapter we will explore how these rights processes unfold in practice. We are concerned with not only the extent to which Indigenous or, also in the case of Norway, non-Indigenous peoples are granted user rights but also the extent to which *divergent relations between landscape and people* (what we have referred to as uncommons) are recognised in the processes and potentially acknowledged as legitimate. Put differently: does the legal process allow multiple world-making landscape practices to exist side by side, or does it crystallise yet another 'one-world world formation'? While most studies of Indigenous land rights and land use focus on the effect of legal land rights processes for Indigenous (potential/alleged) right-holders, we focus instead on the effect of such processes on the kinds of natures/landscapes that are permitted within a legal/bureaucratic format of legal rights claims. What kinds of 'stories' are invited or acknowledged as legitimate statements? To the extent that they are articulated, what are the implications of such divergent articulations for the kinds of futures that are imagined, performed or allowed to unfold? Finally, we discuss the implications of such divergences for ontological multiplicity, what future imaginaries do these legal works produce?

Our focus is on whether local land use practices and relations to land are acknowledged, with a view of not only how different *concepts and practices of nature* are embedded in human landscape practices (compare the Introduction to this book) but also how such different *landscapes and people/citizens/right-holders are performed* or enacted, as sites that allow co-existences of particular relationalities and affordances of both humans and non-humans (Ween 2009; Lien and Davison 2010; Abram and Lien 2011; Ween 2014). We pay attention to how landscapes and people are formatted through the legal process and the entities that are permitted or made legible. We are inspired by Helen Verran's approach to landscapes as people-places (Verran 2002) enacted by particular legal practices as an ethnographic finding, a local emic concept and an outcome rather than an analytical premise.

Our analysis draws on our long-standing ethnographic interest in both north Norway and Australia. Marianne Lien has done fieldwork in Finnmark since the mid-1980s and recently re-engaged old field sites, in relation to questions of domestication. Gro Ween has worked with Indigenous land rights in both countries, with native title claimant groups during her first fieldwork in the Australian north in the late 1990s (Ween 2002) as well as the Finnmark Commission (Pedersen *et al.* 2010). She has been engaged in fieldwork in several Sami areas (Southern Sami, Sea Sami and River Sami). As anthropologists, our fieldwork involves participant observation of significant activities, often with a focus on subsistence activities, but also participation in activities such as political activities, legal and bureaucratic procedures, as well as interviews and conversations, often based upon long-time collaborations, that in Indigenous contexts become necessary extensions of subsistence practices.

Our understanding of nature practices is influenced by Studies of Technology and Science (STS), with a focus on how nature is enacted. Central to such orientations is the understanding that there is always more than one nature, that nature is not out there, not something that only lends itself to one kind of description, and that the ways in which nature is described are always also performative (see, e.g., Latour 1988; Asdal 2008; Abram and Lien 2011). The STS awareness of the agency of documents, legal texts and other bureaucratic devices is well established and has also seeped into anthropological approaches (see, e.g., Riles 2000; Cruikshank 2005; Tsing 2005; Blaser 2010).[3] These indigenous rights cases make apparent that legal texts, propositions, regulations and policy notes have the potential to enforce great change (Ween 2009, 2014). Description of nature, including the act of writing, inscribing in law or bureaucratic regulation in itself, has agency. Legal texts and bureaucratic procedures are in a position to produce an *authoritative narrative*. Historian William Cronon (1992) reminds us to consider the very authority with which such narratives are presented and that this particular vision is achieved by obscuring large proportions of reality: narrative foregrounds and backgrounds to hide discontinuities and contradictory experiences. A powerful narrative constructs common sense, making the contingent seem determined and the artificial seem real (Cronon 1992). It is the work involved in producing the political that is our concern.

We begin by examining the current procedures of the Finnmark Commission against the comparative backdrop of Australian Native Title, highlighting some important differences. We then situate these differences in relation to the colonial/post-colonial context in both nation-states, considering differences as well as similarities in relation to their respective colonial histories, notion of equality and the distinction between settler nations and non-settler nations. Finally, we discuss the implications of such differences in relation to future imaginaries, ontological differences, and suggest that rectifying processes perform not only the allocation of user rights but also a form of ontological politics. But before we turn to the empirical material, let us briefly explain what we mean by (post-)colonial practices and nature practices and how this approach is relevant to our comparison.

Law and (post-)colonial nature practices

While Australia is a settler nation, established through the transition/acquisition of former British colonies in a land already settled by an Indigenous population, Norway's colonial heritage is far more complicated and also less pronounced in the narratives of the nation. Norway is celebrated as a nation that, after 400 years without sovereignty, as an inferior in unions with Denmark and Sweden, finally gained its independence in 1905. In spite of some failed Norwegian efforts in the 1920s/1930s to annex Greenland and the Faroe islands from Denmark, Norway is often distinguished as being relatively 'innocent' in relation to imperial powers in Northern Europe, as it held no colonies overseas. (Colonies were held by the Danish Norwegian Crown, but were not governed from what is known as Norway today.)

The fact that Norway as an independent nation-state (post-1905) has never held colonies overseas does not mean that there were no internal frontiers or settler politics. Swedish/Norwegian nation-building efforts in the nineteenth and twentieth centuries, the imposition of church and state administration on politically strategic regions, and efforts to eradicate Sami religious and cultural practices are trajectories with colonial undertones (Figure 9.1). Such practices reinforced a cultural and political process referred to in Norwegian as *fornorsking* (literally 'Norwegianisation'), which was already ongoing, due to a non-Sami population expansion into Sami areas at a time (mid-nineteenth century) when agriculture was promoted and nomadic practices were oppressed.

Figure 9.1 Sami in front of the early Karasjok church.

With permission of Finnmark fylkesbibliotek (Harald Barbalas postkort-samling).

Framing articulations of land claims: the Australian case

Since we do not have here the opportunity for an extensive comparison of Norwegian and Australian Indigenous rights cases, our comparison will restrict itself to the legal inscription, the practices these instigated, their consequences for the possible relations between people and land, the mapping exercises that these brought and, last but not least, the postcolonial potential of enacting such people-places.

Although land rights activism in Australia started elsewhere than the Torres Strait, decades before this, the story of native title began when the Meriam people represented by the Mabo brothers took their claim to High Court, insisting that they were the traditional native title holders of the Murray Islands and that the Crown's sovereignty should be subject to these rights (Sutton 2003, p. xiv). When the Australian High Court delivered judgement in *Mabo and Ors vs. Queensland* in 1992, it held that native title could be recognised by the common law of Australia. It also stated that Terra Nullius – the constitutional doctrine that Australia was uninhabited at the time of colonisation – was 'discriminatory, unjust and unconscionable' (Sutton 2003, p. xiv). The following year, the Australian government passed the Native Title Act 1993(Cth), creating a statutory regime for protecting and recognising native title. According to this Act, Australian Indigenous people could receive native title provided they were able to prove religious and cultural ties to the land as well as continued use through colonial history. To rectify in this case did not mean overturn, as the title did not affect private ownership rights (Sutton 2003; Smith and Morphy 2007).

In the following years, institutions were established through which Australian Indigenous people could apply to have their rights recognised. A native title claim could be launched in several ways: for example, a group of traditional owners could present their claim to the Native Title Tribunal as an independent act, or in response to an Expression of Interest from 'other interested parties', such as entrepreneurs, state governments, shire councils or claims by other local Indigenous groups. The evidence of traditional owners would be assembled by lawyers, anthropologists and archaeologists, with the knowledge of how to translate evidence of these activities into categories that satisfied the legal criteria of the Act. Anthropologists became particularly important, legally authorized, as they in effect became, to translate and describe, in ways that make sense to the tribunals, knowledge of hunting, fishing and collecting, of sacred sites, ceremonies, artefacts, ancestral myths and legends, of different kinds of family and individual connections with land, including the genealogies of families within each claimant groups (Sullivan 1997; Sutton 2003; Morphy 2007; Strang 2010). Native title cases made it clear that connections between people and land were heterogeneous. In anthropological terms, people's connections to land have many origins, based upon language or family group connections, religious practice, subsistence use or as place of birth. Careful ethnographic work, in collaboration with claimant groups, enabled these rectifying processes to allow the articulation of relations between people and land, radically different from Euro-Australian conventional notions of private property and land use as articulated by settler descendants (Verran 2002; Morphy 2007; Strang 2010).

Framing articulations of land claims: the Norwegian case

The development of a Norwegian land rights struggle gained momentum in the 1970s. A conflict over a proposed hydroelectric dam and power plant on the Alta River (in Finnmark county) led to what has later been known as the Alta-Kautokeino conflict. While the protesters lost and the dam was built, the political conflict contributed to make the Sami visible as an Indigenous population with legitimate interests (Figure 9.2). Coinciding with Indigenous rights developments elsewhere, including the UN, the Norwegian state began a legal and fact-finding process that in the end would recognise the existence of Sami rights. Since the 1990s, the Sami Rights Commissions have requested inquiries into Sami rights positions, land as the foundation for Sami culture, Sami customary law, the rights of Sami south of Finnmark and fishing rights in Finnmark (NOU 1994:21, 1997:4–5, 2001:34, and 2008:5). The production of these official reports went on for almost three decades, primarily based upon written text, historical records, but also analysis of customary law in traditional Sami practices. Although a top-down process, these reports were written by both Sami and non-Sami academics and interest groups, producing what often was ground-breaking new history (Ween 2012a, 2012b). Legal changes brought by the work of the Sami Rights Commission included the Sami Act of 1987, which lay the legal foundation for the Sami Parliament, established in 1989, and the Finnmark Act 2005. This process established Norway as a nation based on the lands of two people (Smith 2004), as well as the recognition of colonial 'wrongs' in relation to the Sami (Ween 2008; Pedersen *et al.* 2010).

Figure 9.2 Surviving the future.

Credit: Liselotte Wajstedt.

The Finnmark Act 2005 transferred 45,000 km^2 or about 96 per cent of the county of Finnmark from the Norwegian state to a land-holding institution, the Finnmark Estate. According to the Finnmark Act, management of the land was to benefit not only the county's population (irrespective of ethnic identification) but also especially Sami culture and reindeer husbandry. It involved the recognition that Sami collectively and individually had acquired rights to land through long-term use of land and water. The Finnmark Act is based on three tiers of rights with different kinds of access to land and its resources (Ravna 2006). Local inhabitants of each municipality[4] have the rights to gather eggs and down, as well as to limited logging on their home land. Inhabitants of the Finnmark County are granted the right to hunt and fish and pick cloudberries in the entire Finnmark Estate, regardless of ethnic identity or municipal residence. The general public also from outside of the county, such as tourists/visitors, have rights to small game hunting, angling[5] and cloudberry picking for subsistence purposes (Ravna 2006, p. 81).

It is explicitly stated in the Finnmark Act that the Sami through times imme-morial have rights to land and waters. However, the Act also acknowledges that other inhabitants in Finnmark have such rights through similar long-term use. This benefits those inhabitants who, for a variety of reasons, including harsh assimila-tion policies, do not identify as Sami.[6] As the self-declared Sami only constitute a majority of the population in the inner parts of Finnmark, the Act was designed to balance Sami interests against the interests of the remainder of the population and to invite cooperation and non-conflict among all ethnic groups in Finnmark.

Following the Finnmark Act and the recommendations of Norwegian Official Reports (NOUs no. 1994:21, 1997:5, 2001:34 and 2008:5), the Finnmark Commission was established with the aims of identifying and facilitating a proce-dure through which rights to land, and the content of such rights, could be established on the basis of customary law and immemorial use (see also Minde 2005, p. 29; Ravna 2006, p. 79). As we will show, the structures instigated to map land rights in Finnmark took a different approach from the Australian native title processes.

Land rights settlements compared

While the legal processes of Indigenous land right settlements bear resem-blances, there are also obvious differences in the ways they are organised, in the kinds of expertise that is engaged and made relevant, the role of potential right-holders and the ways in which potential right-holders are invited to articulate their claims. These differences, we suggest, enable rather different processes of worlding (Tsing 2005) and rather different epistemic commitments to coincide with different potential to articulate particular people-places (Verran 2001). In the following sections, we will address these differences in further detail.

Bottom-up versus top-down

A number of writers have noted the Australian native title legislation and its prac-tices as a legal domestication of the indigenous landscape, in ways that represent new forms of colonisation (Morphy 2007; Weiner 2007). While we acknowledge these apparent shortcomings in the Australian context, we emphasise, nevertheless, the difference between what primarily is a bottom-up approach in the Australian context and the Norwegian top-down approach. In Finnmark, the Commission has initiated a rights-mapping process, area by area. For each specified area, referred to in Norwegian as a *'felt'* (field, which may comprise a few municipalities), a separate timeline is established, a 'window' for articulating claims is announced, and reports are commissioned from research institutions (according to identified needs), chosen after a bidding contest.

While the Australian process was developed with explicit attention to local cultural practices, and at times even with some attention to ontological or epistemic difference,[7] the Norwegian process is run in line with common Norwegian official procedures. When a field is chosen, this is publicly declared in all media. Claims in respect of all kinds of rights (ownership rights or user rights), individual or collective rights, are invited. Claims of rights must include the area for which the claim is provided and also include a short description of the actual historical and legal foundation for the claim (*faktiske historiske og rettslige grunnlaget for kravet*). Open meetings take place, direct communications are initiated with communities and municipalities, such as reindeer-herding siidas,[8] fishermen's organisations. So far, the number of claims for each region has been very limited, and most claims are from larger groups, such as reindeer-herding siidas or local community siidas. Although claims can be articulated orally, it is stated that it is advantageous if the claim is articulated in writing (NIKU Oppdragsrapport 43/2011). Claim forms are two-page documents, stating name, connecting type of right, with an associated explanation. On the form, claimants are asked to tick off whether their claim was for private ownership rights or user rights, individual or collective rights. The form has a small space where claimants can describe the historical continuity that provides the grounds for the claim and the type of claim made (or types of claims). There is little discussion of whether these claim forms are culturally appropriate for Sami or other inhabitants, or if the format allows the articulation of locally perceived relations between people and land. In any case, the number of claims that have been registered is low. Otherwise, funding for the mapping of rights in each region has not been substantial enough to make use of more elaborate fieldwork techniques than interviews. Some are phone interviews, others are formatted as multiple-choice interviews. All claims and the final reports are published on the Finnmark Commission website, including the final deliberative report.

When questionnaires have been sent out to larger audiences in a region, the response rate has also been relatively low. For each region, an 'expert group' is set up with representatives from fishermen, farmers, reindeer herders, Sami organisations, municipal councils and others. Consultant researchers write reports on aspects of the regional subsistence uses, such as uses of the commons, coastal or river fisheries.[9] The final report for each field is written up in a legal terminology by the members of the Finnmark Commission. Here, so far, most of the knowledge referred to is based on historical records (often the same official reports, written by the Sami Rights Commissions that in their time provided the grounds to establish the Finnmark Estate and the Commission). Due to the focus on historical knowledge, the explicit legal terminology and the lack of funding for in-depth fieldwork concerning understandings of subsistence uses, oral accounts of land use, religious use as well as ongoing negotiations over rights over time are narratives backgrounded as a consequence of the legal and text-based framework.[10]

Professional expertise: anthropologists, historians and local experts

In the Australian system, there was significant consideration given to the alterity of the Native Title claimants. As the articulation of a Native Title right relied on an acknowledgement of predominantly oral traditions, it was anthropologists who were granted the most significant role of experts in processes to establish claims (Sutton 2003). Religion was deemed to be essential to the continued nature practices. Religious and ceremonial knowledge was respected in the claimant processes as well as by the courts. The Native Title courts were educated in Aboriginal cultural behaviour, including notions of shame associated with public speaking and the difficulties of male presence in hearings of female religious activities. Knowledge of hunting and gathering was demonstrated in situ. Sites of special importance were documented with significant elders making trips and adding their own stories to GIS locations on maps (Sullivan 1997; Ween 2002; Strang 2010). Family genealogies for entire communities were recorded to ensure that all claimants were heard, to secure knowledge of particular family estates and their boundaries, to find areas where boundaries overlapped and to initiate negotiations between peoples with such overlapping boundaries where it mattered. In this way, oral history became meaningful, not only to anthropologists documenting claims but also for the courts' procedures. In other words, arguably, in Australia, local Indigenous people were able to remain the experts of their own land and knowledge.

In the case of the Norwegian Sami rights process, it was historians and legal practitioners who were given the role of experts. These may or may not be Sami themselves. So far, the Finnmark Commission has been able to make use of prominent experts, many of these Sami, in their mapping exercises. There are also good reasons for this focus on text and history. In Norway, the coexistence of Sami-speaking and Norwegian-speaking people in Finnmark goes back at least 800 years. Consequently, there is a rich and varied collection of historical records. Some of these are oral, some written and some materialise as manifestations in the landscape. At the same time, this focus on written records as source, or truth, in the establishment of factual issues also reflects a more general understanding of land claims as being first and foremost classified historically, and in the legal or judicial realm, rather than in common law. The Finnmark Act is generally acknowledged as a process of setting historical records straight, rather than being about acknowledgement of cultural differences as such. This reflects what Helen Verran has described as the absence of imaginary in Western epistemological traditions (1998, p. 243), one that denies the landscape agency so that 'people own land', while the idea that 'land owns people' is unimaginable; within this order, all space is equivalent (1998, p. 241).

While Australian Native Title claims are mediated through an institutionalised process in which cultural difference is expected to take place (Povinelli 2002; Morphy 2007; Glaskin 2007), in Norway, Sami are expected to be equal citizens, fully equipped and able to express their relations to particular lands or resources in ways that the legal experts will recognise. Next to the preference for written claims, little attention is paid to the difference between local customary law and the Finnmark Commission's legal interpretation. The limited education (often due to World War II) of many potential and real claimants of the older generation is not noted, nor so far in the Commission's work is the tendency of many people, regardless of ethnicity, to be uncomfortable with the style of a legal claim.

Fluidity and ambiguity: the articulation of difference

The acknowledged heterogeneous people-places of Indigenous Australians, accounting for overlapping associations in connection with as many different kinds of associations, such as language group, family group, gender, ritual and subsistence practice, as well as place of birth, are far from the people-places of Sapmi. Text-based or, possibly, interview-based descriptions of people-places provide little in the sense of in-depth descriptions of local nature practices that could have involved, for example, religious practices, practices that in Sapmi have been suppressed or kept secret for centuries (Myrvoll 2011) and in many ways still are. In addition, limited attention is paid to a cultural ethos that values tacit knowledge and, as previously mentioned, the fluidity of social relations (Kramvig 2005) that allows for the acknowledgement of many kinds of identities, kinship and friendships across ethnic divides and subsistence activities. In Sapmi, as in Australia, when individual claims are launched fixing the otherwise fluid and heterogeneous understandings of rights, this often brings tensions and conflict to local communities. This could be one of the reasons why the number of individual or group-based claims in the mapping of fields of rights instigated by the Finnmark Commission has been so low. To take the divergent forms of worlding within Finnmark seriously, we must pay attention to the social dynamics that may unfold in a small community when claims are launched over natures that have been used tacitly by many, in different ways, for centuries.

Concluding notes

In this chapter, we have drawn attention to how landscapes are enacted by particular legal practices. In line with STS-inspired observations of the intervening powers of law and bureaucracy, we have chosen to focus on how the Finnmark Commission articulates Sami and other local nature perceptions as an emic concept, as an outcome. Our interest, furthermore, has been in the opportunities that these enacted people-places offer, in terms of making room both for Sami ontological foundations and epistemological framework and for other forms of uncharted difference entailed in the notion of 'uncommons'.

In the work of the Finnmark Commission, there seems to be little awareness of ontological difference. Not to mention awareness of the many acts of translations involved, in order to successfully navigate the new legal landscape. This new legal intervention creates a number of challenges, such as how to feel comfortable with a legally initiated, top-down procedure; how to articulate nature practices on written forms or in an interview, when such knowledge of nature primarily has been taught on the land and in practice; how to articulate complex forms of coexisting identities or user rights that have evolved in a taken-for-granted manner through decades and that continue to work because they are inherently flexible; and, finally, how to articulate relations to land when such relations of traditional use and intimate belonging are not exclusive, but permit multiple landscape practices and users to co-exist.

Based on evidence gathered so far, particularly of the economic conditions but also, as a consequence, of the methodologies employed in the Norwegian investigations, it is unlikely that there will be room for the inherent fluidity of local long-term co-existence, nor the uncommonness of lands in Finnmark. This is in spite of its well-intended efforts to invite and include locals to articulate and participate in the actual process.

As a consequence, the postcolonial potential of the Finnmark Act as a rectifying practice in a time of transition is limited by the bureaucratic and legalistic framing of the nature and practices involved. The lack of open articulation of ontologically divergent framings of what nature, people and belonging is about might, in our opinion, further hinder the Act's postcolonial potential.

Considering how the format of performing legal rights claims excludes the kinds of relationalities between land and people and living non-humans, and the kinds of 'stories' that are invited to be told and that are silenced, the process of inquiry itself contributes to a flattening of the potentially much more rich and multiple sets of practices that *could* have been made relevant, but were not. It should be noted that, although such multiple ontology is absent from the Finnmark Commission, it is increasingly prevalent in Sami academic traditions (see, e.g., Oskal 1995; Kramvig 2005; Kuokkanen 2006; Sara 2009; Guttorm 2011; Myrvoll 2011; Balto and Østmo 2012). This, we argue, is a colonising process, which is in fact re-colonisation in the name of de-colonisation (for similar perspectives with regard to Australia, see, e.g., Morphy 2007). One could argue that the alterity that has become so institutionalised in the Australian context by some could be equally alienating. It could even be that the Indigenous/non-Indigenous situations co-produced in these nation-states over time are more similar than these two legal framings would indicate.

Acknowledgements

We are grateful to the editors for constructive comments, as well as the support of the Centre for Advanced Study in Oslo, Norway, that funded and hosted the research project 'Arctic domestication in the era of the anthropocene' during the academic year 2015/2016.

Notes

1 Hence, it largely escapes the essentialist trap which is often attributed to and associated with the so-called 'ontological turn'. Rather than positing radically different worlds, we agree with Gad *et al.* who suggest that anthropology must proceed *as if* there are many worlds: 'Rather than making a choice between "multi-culture" and "multi-nature", such studies thrive on the exploration of never-finally-closed nature-cultures; the crystallization of specific ontological formations out of infinitely varied elements' (2015, p. 83).
2 Kvæn is an ethnic minority in Norway descended from Finns who emigrated from the northern parts of Finland and Sweden to northern Norway in the eighteenth and nineteenth centuries. In 1996, Kvæn people were granted minority status, and a few years later, in 2005, the Kvæn language was recognized as a minority language in Norway.
3 For example, in her study of the Fijian preparations to the UN Women's Conference in Beijing, Annelise Riles (2000) was concerned with the artefacts of institutional activity, the objects and subjects of bureaucratic practice, how the practice was conceived and what kind of responses it elicited (p. xiv). In his study of how indigenous work in Chiapas became global through becoming involved in indigenous rights processes, Mario Blaser (2010) was concerned with how modern knowledge practices are performed and how exactly the knowledge involved in such indigenous rights work engages with each other (Ween 2014).
4 A municipality is a governing structure within the county, defined by territory, and with specific responsibilities delegated from the state and with certain autonomy on some political issues (including, for example, whether the official language is Sami, Norwegian or both).
5 Such hunting and angling requires the acquisition of hunting or angling permits, as in the rest of Norway, regardless of citizenship, residence or indigeneity.
6 Intermarriages between Sami, Norwegian, Kvæn and Finnish people have been common in Finnmark. Hence, multiple affiliations and bilingualism have been prevalent, and many refuse to take part in a political and ethnic discourse that defines ethnicity as an either-or, rather than a both-and (for details, see Ween and Lien 2012).
7 For critical perspectives on the politics of difference within Australian native title practices, see Povinelli (2002), Glaskin (2007) and Weiner (2007). For a more hopeful interpretation, see Verran (1998).
8 The Sami traditional unit for social, economic and political organisation.
9 See NIKU reports 42/2011, 43/2011, 241/2011, 172/2013 and NINA report 881/2012.
10 'Felt' rapports written up so far are no. 1 Stjernøya/Seiland, no. 2 Nesseby, no. 3 Sørøya, no. 4 Vararanger halvøya øst and no. 5 Varangerhalvøya vest.

References

Abram, S. and Lien, M.E. 2011. Performing nature at worlds ends. *Ethnos*, 76(1), 3–18.

Asdal, S. 2008. Enacting things through numbers: Taking Nature into accounting. *Geoforum*, 38, 123–132.

Balto, A.M. and Østmo, L. 2012. Multicultural studies from a Sámi perspective: Bridging traditions and challenges in an indigenous setting. *Issues in Educational Research*, 22(1), 1.

Blaser, M. 2010. *Storytelling Globalization from Chaco and Beyond.* Durham, NC: Duke University Press.

Cronon, W.1992. A place for stories: Nature, history, and narrative. *The Journal of American History*, 78(4), 1347–1376.

Cruikshank, J. 2005. *Do Glaciers Listen? Local Knowledge, Colonial Encounters and Social Imagination.* Vancouver, BC: University of British Colombia Press.

de la Cadena, M. 2015. 'Uncommoning Nature' *Apocalypsis, August 22nd 2015—Day 79.* Accessed 24 November 2016. Available at http://supercommunity.e-flux.com/texts/uncommoning-nature/.

Gad, C., Jensen, C. and Winthereik, B. 2015. Practical ontology: Worlds in STS and anthropology. *NatureCulture*, 3, 67–86.

Glaskin, K. 2007. Claim culture and effect: Property relations and the native title process, in *The Social Effect of Native Title: Recognition, Translation and Co-existence*, edited by B.R. Smith and F. Morphy. Canberra, Australia: ANU Press, 79–91.

Guttorm, G. 2011. Árbediehtu (Sami traditional knowledge) – as a concept and in practice, in *Working with Traditional Knowledge*, edited by J. Porsanger and G. Guttorm. Sami Allaskuvla: Diedut, 1, 59–76.

Helen, V. 2001. *Science and African Logic.* Chicago: University of Chicago press.

Kramvig, B. 2005. Fleksible kategorier – fleksible liv [Flexible categories- Flexible Lives]. *Norsk antropologisk tidsskrift*, 16(2–3), 97–108.

Kuokkanen, R. 2006. The logic of the gift: Reclaiming Indigenous Peoples' philosophies, in *Re-Ethnicizing the Mind? Cultural Revival in Contemporary Thought*, edited by T. Botz-Bornstein. Amsterdam and New York: Rodopi, 251–271.

Latour, B. 1988. *The Pasteurization of France.* Harvard, MA: Harvard University Press.

Law, J. and Lin, W.-Y. 2011. Cultivating disconcertment, in *Sociological Routes and Political Roots*, edited by M. Benson and R. Munro. Oxford, England: Wiley-Blackwell, 135–153.

Law of 17th of June 2005, no. 85, *Om lov om rettsforhold og forvaltning av grunn og naturressurser i Finnmark fylke, the Finnmark Act.*

Lien, M.E. and Davison, A. 2010. Roots, rupture and remembrance: The Tasman lives of monterey pine. *Journal of Material Culture*, 15(2), 233–253.

Minde, H. 2005. *The Challenge of Indigenism: The Struggle for Sami Land Rights and Self-Government in Norway.* Delft: Eburon.

Morphy, F. 2007. Performing law: The Yolngu of Blue Mud Bay meet Native Title Process, in *The Social Effect of Native Title: Recognition, Translation and Co-existence*, edited by B.R. Smith and F. Morphy. Canberra, Australia: ANU Press, 59–79.

Myrvoll, M. 2011. *Bare Gudsordet Duger: Om Kontinuitet og Brudd i Samisk Virkelighetsforståelse* [Only God's Word Works: About Continuity and Rupture in Sami Perceptions of Reality]. Doctoral dissertation. University of Tromsø, Norway.

NIKU Oppdragsrapport 42/2011. *Felt1 Stjernøya/Seiland.*

NIKU Oppdragsrapport 43/2011. *Felt 2 Nesseby.*

NIKU Oppdragsrapport 241/2011. *Felt 3 Sørøya.*

NIKU Oppdragsrapport 172/2013. *Felt 6 Varangerhalvøyavest: Berlevåg og Båtsfjord.*

NINA Rapport 881/2012. *Felt 5 Varangerhalvøya øst.*

NOU 1994:21. *Bruk av Land og Vann i Finnmark i Historisk Perspektiv: Bakgrunnsmateriale for Samerettsutvalget* [Use of Land and Water in Finnmark in Historical Perspective: Background Material for the Sami Rights Commission]. Norway: Sami Rights Commission [Samerettsutvalget].

NOU 1997:4. *Naturgrunnlaget for Samisk Kultur* [The Natural Resource Base for Sami Culture]. Norway: Sami Rights Commission [Samerettsutvalget].

NOU 1997:5. *Urfolks Landrettigheter på Bakgrunn av Folkeretten og Utenlandsk Rett, Bakgrunnsmateriale for Samerettsutvalget* [Indigenous Land Rights on the Basis of International Law and Foreign Law, Background Material for The Sami Rights Commission]. Norway: Sami Rights Commission [Samerettsutvalget].

NOU 2001:34. *Samiske Sedvaner og Rettighetsoppfatninger: Bakgrunnsmateriale for Samerettsutvalget* [Sami Customs and Rights Perceptions: Background Material for the Sami Rights Commission]. Norway: Sami Rights Commission [Samerettsutvalget].

NOU 2008:5. *Retten til å Fiske i Havet Utenfor Finnmark: Kystfiskeutvalgets Innstilling* [The Right to Fish in the Coastal Areas Outside Finnmark. The Recommendations of the Coastal Fishing Commission].

Oskal, N. 1995. *Det Rette, det Gode og Reinlykken* [The Right, the Good and Reindeer Luck]. PhD dissertation. University of Tromsø, Norway.

Pedersen, S., Bull, K.S., Ween, G.B. and Solbakken, J.I. 2010. *Sjølaksefiske i Finnmark i et Historisk Perspektiv: Betydningen av Sjølaksefiske for Sjøsamisk Kultur* [Coastal Salmon Fishing in Finnmark in a Historical Perspective. The Importance of Coastal Salmon Fishing for a Coastal Sami Culture]. Report for Finnmarkskommisjonen, Sami Allaskuvla/SEG.

Povinelli, E. 2002. *The Cunning of Recognition. Indigenous Alterity and the Making of Australian Multiculturalism*. Durham, NC: Duke University Press.

Rapport *Felt 1 Stjernøya/Seiland*. Accessed 15 October 2015. Available at https://www.domstol.no/globalassets/upload/finn/rapporter-utredinger-og-kunngjoringer/rapporter/felt-1-rapport---den-trykte.pdf.

Rapport *Felt 2 Nesseby*. Accessed 15 October 2015. Available at https://www.domstol.no/globalassets/upload/finn/rapporter-utredinger-og-kunngjoringer/rapporter/felt-2-rapport.pdf.

Rapport *Felt 3 Sørøya*. Accessed 15 October 2015. Available at https://www.domstol.no/globalassets/upload/finn/rapporter-utredinger-og-kunngjoringer/rapporter/felt-3-rapport.pdf.

Rapport *Felt 5 Varangerhalvøya øst*. Accessed 15 October 2015. Available athttps://www.domstol.no/no/Enkelt-domstol/Finnmarkskommisjonen/Felt-1---3/Avsluttede-felt/.

Ravna, Ø. 2006. *Recognition of Indigenous Peoples' Land Rights Through Modern Legislation: The Case of the Sami People in Norway*. Oslo, Norway: Ministry of Foreign Affairs.

Riles, A. 2000. *The Network Inside Out*. Ann Arbour, MI: University of Michigan Press.

Sara, M.N. 2009. Siida and Traditional Sami Reindeer Herder Knowledge. *The Northern Review*, 30, 153–178.

Smith, B.R. and Morphy, F. 2007. *The Social Effect of Native Title: Recognition, Translation and Co-Existence*. Canberra, Australia: ANU Press.

Smith, C. 2004. *Samerettsutvalget – Tyve år etter* [The Sami Rights Commission – Twenty Years After]. Innledning på Torkel Opsahls Minneseminar, Norsk Senter for Menneskerettigheter.

Stengers, I. 2005. Introductory Notes to an Ecology of Practices. *Cultural Studies Review*, 11(1), 183–196.

Strang, V. 2010. Mapping histories, cultural landscapes and walkabout methods, in *Environmental Social Science Methods and Research Design*, edited by I. Vaccaro, E.A. Smith and C. Aswani. Cambridge, UK: Cambridge University Press, 132–156.

Sullivan, P.J. 1997. *A Sacred Land, A Sovereign People, an Aboriginal Corporation: Prescribed Bodies and the Native Title Act*. Darwin: NARU, ANU.

Sutton, P. 2003. *Native Title in Australia: An Ethnographic Perspective*. Cambridge, UK: Cambridge University Press.

Tsing, A.L. 2005. *Friction: An Ethnography of Global Connection*. Princeton, NJ: Princeton University Press.

Tsing, A.L. 2010. Worlding the Matsutake diaspora: or, can actor–network theory experiment with holism?, in *Experiments in Holism*, edited by T. Otto and N. Bubandt. West Sussex: Wiley and Blackwell, 47–67.

Verran, H. 1998. Re-imagining land ownership in Australia. *Postcolonial Studies*, 1(2), 237–254.

Verran, H. 2002. A postcolonial moment in science studies: Alternative firing regimes of environmental scientists and Aboriginal landowners. *Social Studies of Science*, 32(5/6), 729–762.

Ween, G.B. 2002. *Bran Nue Dae: Indigenous Rights and Political Activism, in Kimberley, Western Australia: A Study of Indigenous Activism*. DPhil dissertation in Social Anthropology. Oxford University, Oxford.

Ween, G.B. 2008. *Indigenous Land Rights in Norway: Does Norwegian Indigenous Land Rights Development have relevance for Africa?* Oslo, Norway: Kartverket.

Ween, G.B. 2009. Blåfjella/Skjækerfjella Nasjonalpark: Om Forvaltning og Utøving av Sted [Blåfjella/Skjækerfjella National Park: About Management and Enactment of Place]. *Norsk Antropologisk Tidsskrift*, (1–2), 95–106.

Ween, G.B. 2012a. Performing indigeneity in human-animal relations, in *Eco-global Crimes: Contemporary and Future Challenges*, edited by R. Ellefsen, G. Larsen and R. Sollund. Farnham, England: Ashgate, 295–312.

Ween, G.B. 2012b. Resisting the imminent death of wild salmon: local knowledge of Tana fishermen in Arctic Norway, in *Fishing People of the North: Cultures, Economies, and Management Responding to Change*, edited by J.P. Johnsen, C. Carothers, K.R. Criddle, C.P. Chambers, P.J. Cullenberg, J.A. Fall, *et al.*. Fairbanks, AK: Alaska Sea Grant,153–171.

Ween, G.B. 2014. Tracking nature inscribed: nature in rights and bureaucratic practice. *Nordic Journal of Science and Technology Studies*, 2(1), 28–34.

Ween, G.B. and Lien, M.E. 2012. Decolonisation in the Arctic? Nature practices and rights in sub-Arctic Norway. *Journal of Rural and Community Development*, 7(1), 93–109.

Weiner, J. 2007. History, oral history and memorization in Native Title, in *The Social Effect of Native Title: Recognition, Translation and Co-Existence*, edited by B.R. Smith and F. Morphy. Canberra, Australia: ANU Press, 215–225.

10 Mining as colonisation

The need for restorative justice and restitution of traditional Sami lands

Rebecca Lawrence and Mattias Åhrén

Introduction

International law has increasingly come to recognise that Indigenous peoples retain rights not only in territories historically and continuously used but also in lands traditionally used but subsequently taken without their consent (e.g. UNDRIP, Art. 28). This legal development is in line with moral and political debates around restorative justice more generally (e.g. Tully 1995; Dodds 1998). Indigenous cultures are intrinsically rooted in their traditional territories,[1] and dispossession of their traditional lands throughout colonisation has had, and continues to have, 'disastrous effects' on Indigenous peoples; it has 'deprived [them] of political sovereignty and has contributed to dislocation and loss of cultural integrity, language, and cultural connection' (Dodds 1998, p. 188). Because of the centrality of lands to Indigenous cultures, international law stipulates that, when previously taken without consent, such lands should ideally be returned through restitution, but when this is not a practical option, compensation in various forms may suffice (e.g. UNDRIP Article 28, CERD General Comment No. 23, para. 5, *Sawhoyamaxa* 128). Settler-states such as America, Canada, Australia and New Zealand, but also Nordic states such as Denmark, have responded in various ways to these developments and debates.

Sweden, however, has not kept abreast of these developments, with regard to the Indigenous Sami people, whose traditional territories span across northern Fennoscandinavia and the Kola Peninsula. This has resulted in repeated criticism from the UN, which has condemned Sweden for failing to recognise and implement the land and resource rights of the Sami in national legislation (for a recent example, see CERD/C/SWE/CO/19-20, para. 17). Some may be surprised that a country that is otherwise known to have a good human rights record has received criticism from the UN for constantly violating the human rights of the Sami people. In this chapter, we argue, however, that Sweden's non-recognition of Sami land and resource rights should in fact not come as a surprise but is rather a logical consequence of Sweden's failure to engage with its colonial past and ongoing present.

In short, Sweden does not see itself as a coloniser. It has long purported the largely rejected 'saltwater' theory (compare Tully 2000, p. 55) in which colonies are defined as territories geographically distant from the imperial 'home'. For instance, a Swedish government official report of 1986 stated that '[i]t is impossible to compare the Swedish influence over the traditional Sami areas with what is ordinarily called colonialism' (Statens offentliga utredningar 1986, p. 164). There have been positive developments since. For instance, Sweden established a Sami Parliament in 1996 and in 2010 gave the Sami constitutional recognition as 'people'. Still, the Sami Parliament lacks significant political mandate (Lawrence and Mörkenstam 2016), and it is also unclear whether the constitutional recognition has otherwise had practical implications. Arguably even more importantly, there has been little development when it comes to the possibility of the Sami to control their traditional territories.

This fact is well illustrated by the position taken by the Swedish government in the so-called *Girjas Case* (ongoing at the time of writing), a highly symbolic and critical case, both for the Swedish state and the Sami, concerning Sami hunting and fishing rights on Crown lands. During the trial, the legal representatives of the Swedish government rejected Girjas Sami reindeer-herding community's assertion that it holds exclusive hunting and fishing rights on its traditional territory, claiming instead that the land was 'a borderless wilderness'; 'a desolate environment, without people or borders'; and 'a no-mans-land', prior to the formation of the Swedish state (field notes, Lawrence, Gällivare District Court, June 2015). These were not isolated statements made by the government's legal representatives on a whim but in fact constituted the very spirit of the government's entire defence against the Girjas Sami community's case, in which the Swedish state in fact questioned the very status of the Sami as Indigenous people. Thus, Sweden mobilises the discourse of *terra nullius* – discussed in more detail below – to reject Sami land claims in the year 2015, while the rest of the world, and the law, has long ago rejected this doctrine as scientifically false and inherently discriminatory.

As a contrasting example, after decades of protests, strikes and rallies by Indigenous people in Australia, Gough Whitlam, then candidate for prime minister, argued in 1972 that his government would 'legislate to give aborigines [sic] land rights – not just because their case is beyond argument, but because all of us as Australians are diminished while the aborigines [sic] are denied their rightful place in this nation' (1972). In other words, the Australian people, according to Whitlam, had a moral obligation to make right an historical wrong. The Aboriginal Lands Rights Act 1976 (Cth) was later passed, at the time effective only in the Northern Territory of Australia, but which nevertheless established representative land councils for traditional owners, mechanisms for the transfer of Aboriginal land back to Traditional Owners, a veto right for Traditional Owners over all mineral exploration and exploitation, and a right for Aboriginal people to share in the benefits of mining activities via royalties.

Later, in the Australian High Court's ruling in *Mabo* (*Mabo v Queensland No. 2*) of 1992, the legal fallacy of *terra nullius* upon the arrival of the British was finally overturned in Australia, and Aboriginal and Torres Strait Islanders were given recognition of their customary rights to land through native title provisions in Australian common law,[2] applicable to the whole country. Our point here is not to present

Australian legal and political responses to Indigenous claims as models.[3] Our aim rather is to draw attention to the importance of a principled recognition of historical injustice and the political will needed to address that history through the return of Indigenous lands to Indigenous peoples and through compensation for lands lost.

In contrast, discussions on restorative justice in a Sami context do not figure in the Swedish political debate. True, the Swedish government has appointed a number of official expert committees (*Statliga offentliga utredningar*) to survey issues addressing or pertaining to Sami land rights, but none of these have led to any political or legislative action. Sweden's position appears largely to be that since, in its view, there has been no colonisation or dispossession, there can be no duty to restore. In essence, this position towards Sami land and land rights could be described as ahistorical. It requires a particular view of the present that simply erases the past. Or, if the past is given any recognition, the passing of time is presumed to invalidate past injustices (Waldron 1992). With this chapter, we therefore seek to argue that history, and time, is relevant. The question of 'land restitution ... brings the past into the present' (Fay and James 2009, p. 1), and by bringing the past to the fore, we seek to challenge the Swedish state's current position on Sami rights.

We argue, by way of developments in international law and theories of restorative justice, that Sami communities (1) retain proprietary interests in lands traditionally used but subsequently taken without consent and (ii) have the right to control access to their respective territories based on rights formally recognised but today not implemented in practice. This includes shares in benefits arising from resource extraction on such lands. The claim is not that the restitution of lands provides a universal solution to all inequalities facing Indigenous peoples (Dodds 1998). We do maintain, however, that it is a critical element in addressing past and ongoing injustices directed at them. Restitution therefore deserves more academic *and* political attention. In a strict legal sense, restitution refers to the return of land to which no formal property right applies, due to previous forms of extinguishment or other acts resulting in the formal legal loss of lands. In this chapter, however, we refer to restitution more broadly as also encompassing the return of lands where formal property rights do *legally* exist but are *politically* not currently recognised.

We begin this chapter by briefly sketching the internal colonisation of Sami lands and how this 'political' colonisation did not necessarily result in non-recognition of Sami rights over land. We subsequently argue that while state recognition of Sami land rights may have diminished throughout the centuries, the rights themselves have never been formally extinguished. We then examine the various arguments for restorative justice, with particular reference to developments in international law. We make our case with specific reference to the Swedish mining industry, although the normative basis for our arguments applies equally to other resource sectors. Mining is only one of many industrial activities that historically have dispossessed, and continue to dispossess, the reindeer-herding Sami of their traditional territories. Yet mining has held, and continues to hold, a particular significance in both Sweden's political economy and its colonial imaginary.

Moreover, while other resource industries operating in the Sami territories have adopted branch-specific voluntary guidelines that call for respect for Indigenous rights, such as the forest industry's Forest Stewardship Council, the Swedish mining industry has not effectively engaged with Indigenous rights issues at a policy level. The Swedish state likewise has failed to streamline its mining and other legislation in accordance with developments in international law. Also, mining provides an illustrative case of the ways in which Sami reindeer-herding communities are left burdened with significant negative effects of resource extraction while being excluded from those benefits ordinarily granted to affected Indigenous communities. By way of example, we demonstrate – counterfactually – the net value of mining royalties currently *not* paid to Sami reindeer-herding communities in Sweden.[4] We conclude that in order for restitution to take place, Sweden must first acknowledge, and critically engage with, both its colonial history and ongoing role in the colonisation of Sami lands.

Sweden's historic (and on-going) colonisation of *Sápmi*

In contrast to external colonialism, in which the colonies and imperial society exist in different territories, internal colonisation refers to the processes whereby 'the colonising [neighbouring] society is built on the territories of the formerly free, and now colonised, peoples' (Tully 2000, p. 39). Sweden's colonial project did not predominantly focus on the new world[5]; instead Sweden turned its gaze to traditional Sami territories in the north. Two interconnected historical trends are particularly relevant to understanding and recognising the way in which the internal colonisation of Sami lands has played out. Acknowledging them is a necessary first step towards recognising the need for restorative justice in the Sami territories.

The first is the close link between the colonisation of Sami territories and Swedish natural resource use. In a Swedish context, the term colonisation may be understood as a process by which the state 'cultivated' the land of the north and put previously 'unused' territories into use (Korpijaakko-Labba 1994, pp. 21–35). The second was increasing *non-recognition* of Sami land rights. Again, Sami land rights were never extinguished but were rather gradually accorded less recognition. This argument is at odds with the state's own narrative of *terra nullius* and challenges historical accounts that tend to conflate political non-recognition with legal extinguishment of rights. We expand upon this below.

Early contacts between the Sami and Scandinavian societies were marked by trade and other forms of peaceful coexistence. True, the Swedish Crown imposed taxes on the Sami population. However, with taxation came the recognition of rights. Sami family groups paid taxes to the Crown for the exclusive right to defined land areas they had traditionally used for reindeer husbandry, hunting, fishing and so forth – so-called Lapp-taxed lands. The system of Lapp-taxed lands paralleled the tax system by which Swedish settlers paid for exclusive rights to their farmed lands. Notably, Sweden's recognition of Sami exclusive rights to lands was spurred by a perception that the inland mountainous and forested areas in Sweden were most effectively utilised through Sami traditional livelihoods (Korpijaakko-Labba 1994, pp. 55–56; Lundmark 2012, p. 36; Päiviö 2012, pp. 13, 66–67, 122–123; Mörner 1982, p. 40; Hansen and Olsen 2007, pp. 292–293). However, as Sweden discovered silver and other natural resources in the Sami traditional territories, an explicit colonial imaginary also began to take shape. 'The North' became known as 'The Land of the Future' (*Framtidslandet*), and silver deposits at *Nasafjäll* were hoped to be 'Sweden's West Indies' (Lindqvist 2007, p. 249).

The discovery of sub-soil natural resources in the Sami territories, however, had no immediate impact on the Crown's position on Sami land rights. Rather, peaceful coexistence largely continued to mark the Swedish–Sami relationship throughout the seventeenth and eighteenth centuries (Korpijaakko-Labba 1994; Päiviö 2012). Still, some reforms took place that arguably would later come to have consequences for the recognition of Sami rights. A 1695 fiscal reform rendered so-called Lapp villages, encapsulating a number of Lapp-taxed lands, the principal Sami fiscal subjects. When the Crown subsequently introduced a comprehensive land rights reform in 1789 (*Förenings-och Säkerhetsakten*) that extended official ownership rights to Swedish holders of farmed taxed lands,[6] the same courtesy was not extended to Sami holders of Lapp-taxed lands. Presumably, a contributing factor to the non-formalisation of Sami land rights was the 1695 fiscal reform. Although the 1695 reform had not abolished the Lapp-taxed lands, the 'collectivised' Sami land use through the Lapp villages,[7] and the corresponding 'depreciation' of the Lapp-taxed lands, may have induced the Crown to view the latter differently compared with Swedish farmed lands (Päiviö 2012, pp. 97–98).

Towards the end of the eighteenth century, Swedish settlement in the Sami territories began to increase and with it competition over natural resources. By the 1800s, Sweden viewed Lapland as its 'depot of raw materials, the equivalent of Africa and India for England' (Baer 1982, p. 14). It increasingly encouraged members of the Swedish population to settle in the 'wilderness [in the North] and thereby expand the [territory of] the Crown'. The view was that 'Lappland [should be brought] into ... the Sweden Empire' (Göthe 1929, cited in Bylund 1956, p. 34). As Swedish settlement in the Sami areas increased, so did conflicts between Sami traditional livelihoods and Swedish settlers (Bylund 1956, p. 33; Lundmark 2006). Moreover, large-scale industrial developments, such as forestry and waterpower, and mining took hold in Sweden, not least in the Sami territories where much of Sweden's resources were situated. With increased Swedish settlement in Sami traditional territories and escalating interest in the natural resources there, 'Sami ownership of land and waters ... became an obstacle in the way of exploitation of the natural resources of Lapland' (Baer 1982, p. 14). Now, Sami land rights, previously perceived beneficial to the Crown, could potentially become a hindrance to Swedish 'development' and interests.

A perception began to emerge that 'uncivilised' Sami land uses, such as reindeer herding, lacked the capacity to 'cultivate' land in manners establishing exclusive rights (Mörner 1982, p. 49; Korpijaakko-Labba 1994; Lantto and Mörkenstam 2008) and must therefore yield to more 'developed' Scandinavian land uses – including resource extraction (Bylund 1956, p. 243). Some of these arguments also took on social Darwinist overtones (Mörner 1982, p. 49; Lundmark 2002; Lantto 2005; Allard 2006, p. 34). Those familiar with that theory recognise here the *terra nullius* doctrine, a common position throughout the colonial world in the late nineteenth century (Tully 2000, p. 40), a discussion to which we shortly return in more detail.

Notably, however, it was the perception of Sami rights, presumably spurred by the 1789 land rights reform as discussed above, rather than the law itself that changed. No attempt was made to officially extinguish Sami land rights, presumably because Swedish legislators and authorities had developed the view that the Sami – by the nature of their society – were simply not in possession of any exclusive rights that could impede Swedish 'development'. Hence, the Crown identified no existing law in need of amendment.[8] As a consequence, whatever land rights Sami communities possessed at the time remained unchanged. The fact that Sweden gradually accorded these rights less and less recognition is from a legal point of view irrelevant.

Theories of Indigenous rights and restorative justice

When international law emerged in Europe, which is generally said to have happened in the second part of the 1600s following the Peace of Westphalia in 1648, it largely did so for the very purpose of facilitating imperialism (Kymlicka 2011, p. 183; Crawford and Koskenniemi 2012, p. 15). The international legal system that surfaced rested heavily on the principle of state sovereignty (as does international law today) and thus authorised the European states to declare Indigenous territories *terra nullius*. The *terra nullius* doctrine has two elements. The first relates to a population's political status, the second to its capacity to establish private rights over land. As to the first element, European sovereigns (and hence international law) held that indigenous societies were uncivilised, as opposed to the civilised European states, and proclaimed that only those populations endowed with civilisation could be subjects of international law and thus hold sovereign – that is, *political rights* – over their territories (Cassese 2005, p. 28; Koskenniemi 2005, p. 152; Crawford 2006, pp. 263–267).

The second element professes that in order to establish *private rights* over land, one must improve on, that is, add value to it. British philosopher John Locke (1632–1704), generally identified as a founding father of this line of thought, posited that uncultivated land cannot constitute property. Rather, man must render the land valuable and productive to claim rights thereto (Locke 1689, pp. 309, 312–315; Tully 1995, p. 72; Gilbert 2006, pp. 24–26). Accordingly, the *terra nullius* doctrine denied Indigenous peoples private rights over land as well, as their land uses had not sufficiently impacted on it – that is, they had not made the land sufficiently 'European-like'.

In summary, a 'dynamic of difference' argument was invoked to conclude that colonisation of Indigenous territories was legal (Anghie 1999, pp. 24–25). Deemed politically and culturally inferior, Indigenous peoples were considered 'undeserving of equal consideration' (Ivison 2003, p. 335) and could, from a legal perspective, be described as 'mere ghosts in their own landscapes' (Huff 2005, p. 298). The view that Indigenous landscapes are, legally speaking, empty, irrespective of actually being inhabited, became entrenched during the centuries that followed and was readily adopted by the contemporary international legal system that took form after the establishment of the UN. As Jérémie Gilbert has argued, 'whereas early international law was aimed at providing the colonizers with a legacy for their conquest, modern international law aimed at justifying the stability of these conquests for either the colonizers or their descendant States' (2006, p. 21). The *terra nullius* doctrine was thus an integral part of international law for more than three centuries. It has, however, been seriously challenged by the contemporary Indigenous rights discourse.

Contemporary international law recognises the political status and rights of Indigenous peoples primarily through the acknowledgement that they are beneficiaries of the right to self-determination, as enshrined in Article 3 of the UN Declaration on the Rights of Indigenous Peoples (UNDRIP), for example. (On the relevance of the right to self-determination of Indigenous peoples, see generally Åhrén 2016, chapters 5 and 6, with references.) At present, international legal sources offer limited guidance as to the scope and content of the right to self-determination when applied to Indigenous peoples. Still, given the centrality of lands and natural resources to the ways of life and cultural identity of Indigenous peoples, that right's resource dimension must reasonably be relatively far-reaching when applied to such peoples. If not, the right would be essentially meaningless to them. This conclusion finds support in UNDRIP Article 32, which proclaims that states must obtain the consent of Indigenous peoples prior to approving resource extraction on their lands.

As to the notion that the nature of land uses of Indigenous peoples disqualifies them from establishing private rights thereto, this position has been challenged by the inclusion of the right to equality into the contemporary international legal system. A legal system that rests heavily on the principle of equality cannot reasonably claim that certain segments of society are per se incapable of establishing rights over land. Consequently, soon after the creation of the UN, courts around the world rejected the *terra nullius* doctrine as inherently discriminatory (Åhrén 2016, section 8.3, with references). More recently, international law has gone beyond 'merely' calling for formal equality. The right to non-discrimination no longer only requires that equal situations be treated equally. It also obliges states to treat differently those whose situation is significantly different compared with the majority population (*Thilmmenos*). In an Indigenous land rights context, this means that domestic courts and other bodies must culturally adjust the criteria necessary to establish property rights over land through use. Indigenous communities no longer need to have used lands in ways common to the majority culture in order to establish property rights thereto. Rather, international law now requires states to recognise that Indigenous communities hold property rights to lands traditionally used in manners common to their particular cultures (Åhrén 2016, section 8.5, with references). Recall also that land rights interests of Indigenous communities need not be restricted to territories traditionally and *continuously* used. An increasing number of international legal sources provides that such communities retain property rights interests also in territories traditionally used, but subsequently taken without consent (see, e.g., UNDRIP Article 28).

These developments in international law have been matched in Swedish domestic jurisprudence. In the *Lapp Taxed Mountains Case* (*Skattefjällsmålet*), the Supreme Court confirmed in principle that Sami reindeer-herding communities pursue reindeer herding, hunting and fishing based on property rights established through traditional use, thus rejecting the government's argument that reindeer herding is pursued based on a state privilege.[9] Clearly, this argument can be viewed as having its roots in the theory that land uses common to the Sami culture can per se not result in rights, as discussed above.

In the subsequent *Nordmaling Case* (*Nordmalingsmålet*), the Supreme Court responded to the more recently articulated developments in international law and adjusted the criteria necessary to establish property rights to land through traditional use to the particularities of Sami reindeer herding. The Court proclaimed that Sami reindeer-herding communities that have pursued reindeer herding in manners common to the Sami culture have established property rights although the land uses had not resulted in rights if measured against standards set by Scandinavian agrarian culture. Most recently, in the ongoing Girjas case (mentioned in the introduction), the Gällivare District Court acknowledged in their ruling (Mål nr. T 323-09) that the Girjas Sami community has an exclusive right to hunt and fish on their traditional lands, without interference by the Swedish state, and that the community also has the right to temporarily transfer those rights to others. The District Court's decision is in line with a more general trend both in international law and Swedish domestic jurisprudence: to recognise Indigenous peoples' property rights established through traditional use. However, the Swedish state has appealed the District Court's decision, illustrating well the disjuncture between developments in international law and Swedish domestic jurisprudence on the one hand and the Swedish state's position on the other.

As a general rule, the right to property mandates the holder of the right to deny others access to the property. Naturally, this aspect of the right applies also in an Indigenous context, as anything else would be discriminatory (Åhrén 2016, section 9.2, with references). Consequently, resource extraction should, from a legal perspective, normally only occur on the traditional territories of Sami reindeer-herding communities if they give their consent. The extent to which this norm is adhered to in practice is a different matter, however.

The Swedish mining industry and Sami reindeer herding

Background

Sweden has an old mining industry, which has held, and continues to hold, a particular significance for both its political economy and colonial imaginary. 'Norrland' – the northern region of Sweden – still appears at times to be seen by Swedish politicians as a 'colony' with an infinite depot of raw materials. For instance, during a speech at a major mining conference in Sweden in 2014, the then Swedish Ambassador for the Arctic encouraged foreign and domestic mining companies to appreciate the mining potential of northern Sweden. In doing so, he quoted the famous Swedish statesman Axel Oxenstierna from the 1600s: 'Oxenstierna said we have an India in Norrland, we need to take advantage of that. And I believe this is still true, he got it right' (conference field notes, Lawrence, 'The Mining and Mineral Industry of the Future', Grand Hotel, Stockholm, 29 January 2014).

In line with this view of the north as a colony, Swedish politicians, and the population in general, largely hold on to the notion that it is a self-evident fact that mines generate great wealth, bring local employment and that there are no viable industries that can replace mining in the northern peripheries (Lööf 2013). However, Sweden's mining industry constitutes around only 1 per cent of its GDP, relatively low in comparison to other traditional mining countries such as Australia with a mining production of approximately 9 per cent of its GDP. As to employment, mining has become an increasingly capital-intensive and low labour-intensive industry. It now provides far fewer jobs than it did historically, both relative to production levels and in absolute terms (SGU 2013), and the fly-in-fly-out phenomenon has contributed to an increased reliance on non-local labour.

In 2012,[10] 12 of Sweden's 16 active metal mines, and 98.5 per cent of the value of the mineral extraction, were situated on Sami traditional territories.[11] A similar proportion of the exploration for new mines also occurred in Sami areas. Because of the impacts exploration and mining have on traditional Sami reindeer herding, something we discuss in more detail below, Sami protests over increased mining activities on their lands have been on the rise. In this context, Swedish interests often reduce Sami rights claims to those of 'special interest groups' (see Howitt & Lawrence 2008, p. 96; Mörkenstam, 1999) framed as thwarting not only mining activities per se but regional development, the national interest and even the Swedish state's moral and political obligation to supply the world, and in particular the EU, with minerals. Yet the Sami have little real influence over if and how mining and exploration takes place on their lands, despite the rights arguments articulated above. Thus, in this sense, mining in Sweden remains very much a colonial practice.

The framing of rights as 'interests'

When a mine is established on land to which a property right applies, the infringement of that right must meet certain criteria, in order for it to be lawful. For instance, the legitimate aim criteria require that the infringement is motivated by a legitimate and substantial societal need (Additional Protocol 1 to the European Convention on Human Rights, Article 1). For the proportionality criterion to be met, the infringement must strike a fair balance between the interest of society as a whole in a mine and the interest of the property right holder not to have the land damaged or taken away. The idea is that there are limitations as to what sacrifices one can lawfully request the few to make for the benefit of the many. The infringement must not leave the property right holder with a disproportionate and excessive burden (*Draon v France* 2005; *James and others v United Kingdom* 1986).

Swedish mining law, however, simply assumes that the legitimate aim and proportionality criteria are met in cases of mining, including in relationship to Sami reindeer-herding communities. Mining concessions are granted through an administrative and political process. The Sami reindeer-herding community has no recourse to a court of law that can examine whether the mine violates its property rights, irrespective of the damage caused. Again, this would appear to be a reflection of the assumption that Sami livelihoods can have no legal rights relevant to Swedish 'development': Swedish colonialism proceeds in the form of modern extractive activities with no regard to Sami rights.

One can seriously question the legality of simply assuming that the legitimate aim and proportionality criteria are automatically met, also with regard to Indigenous peoples' traditional livelihoods. For instance, the UN Special Rapporteur on the Rights of Indigenous Peoples (SRIP) has concluded that '[a legitimate aim] is not found in mere commercial interests or revenue-raising objectives, and certainly not when benefits from the extractive activities are primarily for private gain' and further that '[the proportionality criterion] will generally be difficult to meet for extractive industries that are carried out within the territories of indigenous peoples without their consent' (UN 2013, pp. 35–36).

One may reasonably infer that the Special Rapporteur's conclusions are largely reflective of international law (Åhrén 2015, section 9.2, with references). Notwithstanding, rather than allow the relationship between Swedish mining and Sami reindeer herding to be settled by *rights*, Sweden appears anxious to frame this debate in terms of two 'competing' *interests*. Or to be precise, the idea seems to be that these interests are in fact not at all competing, but can rather peacefully coexist.[12] In our view, this optimistic view comes largely from three false assumptions.

The first assumption is that the traditional territories of the Sami reindeer-herding communities are so vast that they can necessarily sustain mining activities. For example, politicians and mining representatives tend to argue that reindeer herding takes place on around 50 per cent of Sweden's land mass, whereas mining concessions are limited to less than 0.05 per cent (Weihed and Ahl 2012). Yet this observation does not take into account the reindeer-herding cycle. Reindeer graze in different pasture areas during different seasons. Each reindeer-herding community must have access to sufficient summer, winter, autumn, and spring pastures as well as to unblocked migration routes and resting areas in between, in order for the reindeer to survive. For these reasons, a mine that 'only' consumes a 'smaller' percentage of a reindeer-herding community's grazing land can have detrimental effects. To illustrate, a comparison can be made with a two-storey house. If one takes away a few stairs out of the staircase between the floors, this arguably only deprives the house-owner of a small percentage of the total floor area. Yet it renders the entire second floor inaccessible.

Second, every new development is assessed in isolation and the Sami areas are thus presumed to be free from other intrusions. Both the mining industry and Swedish authorities regularly overlook, or at least downplay, the cumulative impacts of already existing mining operations, as well as of other activities such as forestry, waterpower, infrastructure, tourism, and windmills. At the same time, Sami reindeer-herding communities regularly state that they are at breaking point and cannot sustain further loss of pasture land (Lawrence and Kløcker Larsen, forthcoming).

Finally, mitigation measures are presumed to ameliorate all impacts on Sami reindeer herding. Yet many mining impacts cannot be mitigated, such as permanent loss of vital pasture areas and the severing of migration routes. Moreover, mitigation measures such as increased use of artificial fencing, supplementary feeding and transport of reindeer by trucks rather than natural migration fundamentally change the nature and viability of traditional Sami reindeer herding. Such measures thereby undermine a cultural practice, which in turn constitutes a threat to the Sami identity and culture (Howitt and Lawrence 2008; Lawrence and Kløcker Larsen, forthcoming). Mitigation measures forced on Sami reindeer-herding communities also easily become part of a self-reinforcing logic articulated by the mining industry: the fact that some reindeer herding-communities already depend on measures such as supplementary feeding and truck transportations of reindeer indicates not only that they *can* adapt but that they *have* already done so without detrimental impacts on Sami reindeer herding or the Sami culture in general. This ignores, however, the fact that Sami reindeer-herding communities have not adopted these practices freely or without damage caused to them. Rather, they have been forced to do so against the backdrop of already existing developments, disturbances, loss of pasture areas and, not least, because of the prevailing view that Sami land uses are simply inferior to the Western 'cultivation' of land through development and should therefore give way (Lawrence 2014).

By relying on these three erroneous assumptions, the Swedish government's mantra in response to any assertion that mining causes a threat to the Sami reindeer-herding culture is that mining and reindeer herding can harmoniously coexist. Swedish authorities responsible for issuing mining permits reflect the same sentiment in their decisions where they commonly take for granted that coexistence is almost always possible.

Lax regulation

The harm that mining can potentially cause to the Sami culture is amplified by Sweden's aspiration that the mining industry should further expand in the Sami territories. International mining companies view Sweden as one of the most mining-'friendly' countries in the world, based on perceptions of environmental regulation, investment security, lack of political risks and essentially non-existent royalty fees (Wilson and Cervantes 2014). Sweden is unique, in so far as it does not have a royalty system comparable to other countries. Mining companies are required to pay a symbolic 'mineral compensation' (*mineralersättning*, in Swedish) of 0.2 per cent[13] *ad valorem* (i.e. the value of extracted ore). This is calculated the same way as an *ad valorem* royalty fee would be, although it is so low – in comparison to other mining countries – that Sweden is generally regarded as simply not having a royalty system (see, e.g., Otto *et al.* 2004, p. 275).

Swedish politicians and civil servants commonly argue that the low level of mineral compensation is justifiable given that Sweden, in their view, has the most stringent environmental regulation in the world (Lööf 2013). Such claims, however, appear to sit poorly with empirical facts. For example, in 2014, an investigation revealed that it would cost Swedish tax-payers over SEK 200 million (Euro 20 million) to clean up environmental pollution at the abandoned mine sites at Svartträsk and Ersmarksberget and the adjoining processing plant at Blaiken. During production, the responsible mining companies had set aside only SEK three million (Euro 300,000) in security for mine site rehabilitation and then filed for bankruptcy (Müller 2012). For the affected Sami communities, the mine sites and remaining associated pollution left impacted grazing lands damaged and unusable. Because of these cases, and others, the Swedish National Audit Office is undertaking an investigation of the current Swedish legislation for mine rehabilitation, financial security and monitoring (Riksrevisionen 2015). However, several Sami reindeer-herding communities have also been left without compensation for damages caused by mining after mining companies have gone bankrupt,[14] and this kind of problem is not currently under any formal investigation.

All drawbacks, few benefits

As touched upon, while Sami reindeer-herding communities must carry many of the negative impacts of mining, most of the positive aspects are streamed to other sectors of society. This is in contrast to general international standards among industrialised countries that host Indigenous peoples, according to which benefits are shared with them when industrial projects such as mining are carried out on their lands.

The scale of the Sami 'loss' can be illustrated by a calculation for one year. In 2012, mines situated on Sami traditional land in Sweden generated a total production value of around 4.3 billion Euro.[15] While the Sami currently receive no royalties,[16] in other industrialised states with Indigenous peoples, it is basically standard that such peoples receive a royalty rate between 2 and 3 per cent *ad valorem* for mining on their lands (O'Faircheallaigh and Gibson 2012, p. 11; SRIP Report 2013). Consequently, the royalties the Sami *did not* receive in 2012 due to Sweden *not* applying international norms can be calculated at between Euro 85 million (at 2 per cent *ad valorem*) and Euro 128 million (at 3 per cent *ad valorem*). Thus, one can say that Sweden 'deprives' the Sami of roughly 100 million Euro annually, simply by not applying the same standard as other comparable countries. Note that this example highlights the amount owed to the Sami by the mining sector for one year. A genuine discussion of restorative justice naturally needs to look both back in time, over all the years of lost benefits to the Sami and across all resource sectors.

Conclusions

With this chapter we seek to provoke a debate. We have counterfactually estimated the mining royalties due to the Sami because there is a larger point to be
made. The point is not that monetary compensation can make amends for past
injustices. Recall that under international law, compensation in the form of return
of land is the norm and preferred option. Monetary compensation is only valid in
those cases where lands can no longer be returned, quite simply because it is inad-
equate in addressing the injustice faced by Indigenous communities dispossessed
of their traditional lands. Indigenous peoples' cultures, and their very existence,
are intrinsically linked to their continued access to lands and natural resources
traditionally used (Scheinin 2000; Gilbert 2006, p. 148).

As mentioned, Sweden did historically recognise exclusive Sami rights over
territories traditionally used. But as Sweden commenced its industrialisation in
the late 1800s, its interest in natural resources in the Sami territories grew and
its recognition of Sami territorial rights correspondingly diminished. Sweden's
invocation of the *terra nullius* rhetoric was used to justify its right to access
the Sami territories for industrial purposes, thus echoing Western liberal arguments used elsewhere throughout the colonial world: Indigenous people did
not cultivate the land and were hence unable to acquire property rights to them.
'Uncivilised' Sami nomadic reindeer herding and other land uses could thus not
reasonably result in rights valid in competition with more 'developed' natural
resource extraction.

Somewhat alarmingly, the discourse of *terra nullius* is not, however, limited to
a distant, colonial past. Judging by the position taken by the Swedish government
in the *Girjas Case,* as referred to above, and decisions taken by it and Swedish
authorities in mining permit issues, Sweden's view remains largely the same as in
the 1800s. The Swedish state's de facto position is that Sami land uses cannot – by
their very nature – result in exclusive rights relevant to mining and other industrial
activities. This position has become all the more outdated given the developments
both in international law but also in Swedish domestic jurisprudence. It is against
this background we argue that colonialism is not only a chapter in Sweden's past.
It is very much part of the present.

But we have also argued that the changed attitude towards Sami land rights
throughout the centuries did not actually formally diminish those rights. Rather,
they were rendered invisible. Here, Deborah Bird Rose's metaphor of a hall of
mirrors is instructive: the state uses circular and self-legitimating Eurocentric
knowledges that only recognise themselves and render other – Indigenous –
knowledges invisible (Howitt and Suchet-Pearson 2003, p. 558). Thus one could
argue that the Swedish state:

> sees itself within a hall of mirrors; it mistakes its reflection for the world, sees
> its own reflection endlessly, talks endlessly to itself, and, not surprisingly,
> finds continual verification of itself and its worldview.
>
> (Rose 1999, p. 177)

In order to break out of this 'monologue masquerading as conversation' (Rose 1999. p. 177), the Swedish state would need to push its gaze past its own reflection and back into the space of time. It would require a deep and serious engagement with historical injustices by acknowledging its colonial past. But it would also require recognising that those same historical injustices are a part of the ongoing present and are perpetuated and exacerbated by the continuing colonisation of Sami lands through modern-day resource activities. Finally, it would also require an acknowledgement that the ongoing colonisation of the Sami traditional territories constitutes a threat to the existence of the Sami culture and people. In this context, the Swedish state has an urgent obligation to act not only morally but also according to international law.

Acknowledgements

We would like to thank the following people for their critical and constructive comments during the writing process: Christina Allard, Lars-Anders Baer, Malin Brännström, Aaron Maltais, Ulf Mörkenstam, Ciaran O'Faircheallaigh, Håkan Tarras-Walberg and the editors. We would also like to acknowledge the valuable assistance of Stefan Tuoma at the Mining Inspectorate of Sweden in calculating mineral production values for 2012. The arguments contained within this chapter remain the sole responsibility of the authors.

Notes

1 Consequently, from an international legal perspective, it is precisely the intrinsic connection to a particular territory that renders a population an Indigenous people (and not an ethnic minority) (UN Doc. E/CN.4/Sub.2/1983/21/Add.8).
2 Since these changes, Indigenous people have won recognition of native title rights to nearly half of the Kimberley region, in most cases as 'exclusive possession' native title, the strongest form available under the native title legislation. It is, however, the case that Indigenous people in 'settled' Australia (i.e. NSW and Victoria) have faced difficulties in establishing their native title rights.
3 See, for example, Povinelli (2002) for a critique of Aboriginal land rights in Australia.
4 While in Sweden the state has not formally declared whether it is the Crown or the land-owner who is the official owner of concession minerals, this has no bearing on our arguments concerning indigenous property rights established through traditional use or the right to restitution.
5 That said, Sweden does have a colonial history also in the new world. It was, however, brief and the territories small in size, in comparison with other major colonial powers at the time.
6 There were "Lapp taxed lands", held by reindeer herding Sami, and there were "farmed taxed lands", held by settlers/ farmers.
7 Sami land use had begun to change. Relatively stationary small-scale reindeer husbandry with domesticated reindeer kept close to the household gradually gave way to larger-scale reindeer herding where semi-domesticated herds migrated over vast areas. To manage the larger herds, Lapp-taxed lands had to cooperate, depreciating the practical relevance of these lands (Lundmark 2012, p. 32) in favour of the Lapp villages.
8 For instance, as far as Sami customary rights were concerned, the first Reindeer Grazing Act of 1886 clearly intended not to change, but rather to codify, already existing Sami land rights (Bäärnhielm 2005, pp. 70–71).

9　See Mörkenstam (1999) for an account of the construction of Sami rights as 'state privileges' in Swedish political discourse.

10　We have used figures from 2012 because at the time of drafting this chapter in 2014 they were the most recent. Note, however, that 2012 was a particularly profitable year for the Swedish mining industry, and that by 2015 iron-ore prices had dropped to around half of what they were in 2012.

11　Of the 16 mineral mines operating during 2012, only four were located in the southern half of Sweden and therefore not on traditional Sami lands (i.e. Garpenberg mine, Danemora mine, Lovisagruvan and Zinkgruvan). These four mines are relatively small-scale in production value and accounted for only 1.5 per cent of the total production value of extracted minerals in Sweden for 2012.

12　For discussion of a similar discourse of coexistence in the forestry and wind power industries, see Lawrence (2007, 2014), respectively.

13　Only those mining concessions granted from 2005 are obliged to pay mineral compensation (*mineralersättning*, in Swedish). Twenty five per cent of the mineral compensation is paid to the Swedish state and 75 per cent is paid to the owner of the land; see Chapter 7 of the Swedish Minerals Act, Minerallagen SFS 1991: 45. Sami reindeer-herding communities receive no share of the mineral compensation.

14　For example, Muonio Sami reindeer-herding community has a compensation agreement with the mining company Northland Resources to cover, among other things, costs associated with the loss of grazing pastures due to the mine site. However, Northland Resources is now bankrupt and the Muonio Sami reindeer-herding community has been left with damaged grazing lands and no compensation to cover ongoing costs.

15　In 2012, the total production value of mineral extraction in Sweden was approximately 4.33 billion Euro. This figure was estimated in consultation with the Swedish Minerals Inspectorate per email and telephone during February 2014 and based on statistics in the report *Bergverksstatistik*, 2012 (SGU 2013). Ninety-eight-and-a-half per cent of this production value took place on traditional Sami lands (see footnote 11), equalling 4.26 billion Euro.

16　Instead of receiving royalties in accordance with international norms, affected Sami reindeer-herding communities normally receive very limited compensation for direct damages pertaining to the immediate loss of pasture lands. In many cases, Sami reindeer-herding communities find that the indirect costs caused by mining operations – in terms of cumulative impacts and secondary impacts such as disturbances caused by noise, dust and traffic – are not compensated for. Moreover, much of the damage cannot be measured in monetary terms. See Lawrence and Kløcker Larsen (forthcoming) for a discussion (in Swedish) of Gällivare skogssameby reindeer-herding community's experience of the Boliden Aitik mine.

References

Åhrén, M. 2015. To what extent can indigenous territories be expropriated, in *Autonomous Sámi Law: Indigenous Rights in Scandinavia*, edited by C. Allard and S. Funderud Skogvang. Farnham, UK: Ashgate.

Åhrén, M. 2016. *Indigenous Peoples' Status under the International Legal System*. Oxford, UK: Oxford University Press.

Allard, C. 2006. *Two Sides of the Coin: Rights and Duties, The Interface between Environmental Law and Saami Law based on a comparison with Aoteoaroa/New Zealand and Canada*. PhD thesis. Luleå University of Technology, Sweden.

Anghie, A. 1999. Finding the peripheries: Sovereignty and colonialism in nineteenth-century international law. *Harvard International Law Journal*, 40, 1–80.

Bäärnhielm, M. 2005. Jakt och fiskerätten i renskötselområdet, in *Vem får jaga och fiska? Rätt till jakt och fiske i lappmarkerna och på renbetesfjällen*. Stockholm: SOU, 17, 59–100.

Baer, L.A. 1982. The Sami – An indigenous people in their own land, in *The Sami: National Minority in Sweden*. Rättsfonden Stockholm: Almqvist & Wiksell International.

Bylund, E. 1956. *Koloniseringen av Pite Lappmark t.o.m år 1867*. PhD thesis. Uppsala University, Almqvist & Wiksells Boktryckeri AB.

Cassese, A. 2005. *International Law*. Oxford, UK: Oxford University Press.

Crawford, J. 2006. *The Creation of States*. Oxford, UK: Clarendon Press.

Crawford, J. and Koskenniemi, M. 2012. *International Law*. Cambridge, UK: Cambridge University Press.

Dodds, S. 1998. Justice and indigenous land rights. *Inquiry: An Interdisciplinary Journal of Philosophy*, 41(2), 187–205.

Draon v France. (2005) 42 EHRR 78 (ruling by the European Court on Human Rights).

Fay, D. and James, D. (eds). 2009. Restoring what was ours: An introduction, in *The Rights and Wrongs of Land Restitution: Restoring What Was Ours*. Abingdon, UK: Routledge-Cavendish, pp. 1–23.

Gilbert, J. 2006. *Indigenous Peoples' Land Rights under International Law*. Ardslay, NY: Transnational Publishers.

Hansen, L.I. and Olsen, B. 2007. *Samenes historie fram til 1750*. Oslo, Norway: Cappelen Akademisk Førlag.

Howitt, R. and Lawrence, R. 2008. Corporate culture and indigenous peoples, in *Earth Matters: Indigenous Peoples, Corporate Social Responsibility and Resource Development*, edited by C. O'Faircheallaigh and S. Ali. Sheffield, UK: Greenleaf, pp. 83–103.

Howitt, R. and Suchet-Pearson, S. 2003. Ontological pluralism in contested cultural landscapes, in *Handbook of Cultural Geography*, edited by K. Anderson, M. Domosh, S. Pile and N. Thrift. London, UK: Sage, 557–569.

Huff, A. 2005. Indigenous land rights and the new self-determination in 16 Colorado. *Journal of International Environmental Law and Policy*, 295.

Ivison, D. 2003. The logic of aboriginal rights. *Ethnicities*, 3(3), 321–344.

James and others v United Kingdom (1986) 8 EHRR 123 (ruling by the European Court on Human Rights).

Korpijaakko-Labba, K. 1994. *Om Samernas rättsliga ställning i Sverige-Finland: en rättshistorisk utredning om markanvändningsförhållanden och rättigheter i Västerbottens lappmark före mitten av 1700-talet*. Helsinki, Finland: Lakimiesliiton Kustannus.

Koskenniemi, M. 2005. *From Apology to Utopia*. Cambridge, UK: Cambridge University Press.

Kymlicka, W. 2011. Beyond the indigenous/minority dichotomy? in *Reflections on the UN Declaration on the Rights of Indigenous Peoples*, edited by S. Allen and A. Xanthaki. Oxford, UK: Hart Publishing, 1–24.

Lantto, P. 2005. Förmyndare för de 'fria naturbarnen': Lappväsendet och svensk samepolitik 1885–1971. *Thule, Kungliga Skytteanska Samfundets Årsbok 2005*.

Lantto, P. and Mörkenstam, U. 2008. Sami rights and Sami challenges. *Scandinavian Journal of History*, 33(1), 26–51.

Lawrence, R. 2007. Corporate social responsibility, supply-chains and Saami claims: Tracing the political in the Finnish Forestry Industry. *Geographical Research*, 45(2), 167–176.

Lawrence, R. 2014. Internal colonisation and indigenous resource sovereignty: Wind power developments on traditional Saami lands. *Environment and Planning D: Society and Space*, 32(6), 1036–1053.

Lawrence, R. and Kløcker Larsen, R. Forthcoming. Då är det inte renskötsel: Konsekvenser av en gruvetablering i Laver, Älvsbyn, för Semisjaur Njarg sameby. Research Report, Stockholm Environment Institute.

Lawrence, R. and Mörkenstam, U. 2016. Indigenous self-determination through a government agency? The impossible task of the Swedish *Sámediggi* (Sámi Parliament). *International Journal of Minority and Group Rights*, 23(1), 105–127.

Lindqvist, A. 2007. *Jorden åt folket. Nationalföreningen mot emigrationen 1907–1925.* PhD thesis. Department of Historical Studies, Umeå University, Sweden.

Locke, J. 1689. *Two Treatises of Government.* London, UK: Thomas Tegg.

Lundmark, L. 2002. *'Lappen är ombytlig, ostadig och obekväm', Svenska statens samepolitik i rasismens tidevarv.* Umeå Sweden: Norrbottensakademiens skriftserier nr 3.

Lundmark, L. 2006. *Samernas skatteland i Norr och Västerbotten under 300 år.* Stockholm: Rönnels Antikvariat.

Lundmark, L. 2012. *Stulet Land.* Stockholm: Ordfront Förlag.

Lööf, A. 2013. Höga miljökrav på gruvor motiverar låg mineralavgift. *Dagens nyheter*, 2 October. Accessed 2 October 2013. Available at http://www.dn.se/debatt/hoga-miljokrav-pa-gruvor-motiverar-lag-mineralavgift/.

Mörkenstam, U. 1999. *Om 'Lapparnes privilegier'. Föreställningar om samiskhet i svensk samepolitik 1883–1997.* PhD thesis. Stockholm Studies in Politics 67, Stockholms universitet, Statsvetenskapliga institutionen.

Mörner, M. 1982. The land rights of the Sami and the Indians – A historical comparison in the supreme court, in *The Sami: National Minority in Sweden*, edited by B. Jahreskog. Rättsfonden: Almqvist & Wiksell International, 36–61.

Müller, A. 2012. Gruvsanering – dyr smäll för skattebetalarna. *SVT Nyheter Västerbotten*, 9 December. Accessed 1 September 2015 Available at http://www.svt.se/nyheter/regionalt/vasterbotten/gruvsanering-dyr-small-for-skattebetalarna.

O'Faircheallaigh, C. and Gibson, G. 2012. Economic risk and mineral taxation on Indigenous lands. *Resources Policy*, 37(1), 10–18.

Otto, J., Andrews, C., Cawood, F., Doggett, M., Guj, P., Stermole, F., *et al.* 2004. *Mining Royalties: A Global Study of Their Impact on Investors, Government, and Civil Society.* Washington, DC: The International Bank for Reconstruction and Development/The World Bank.

Povinelli, E. A. 2002. *The Cunning of Recognition: Indigenous Alterities and the Making of Australian Multiculturalism.* Durham, NC: Duke University Press.

Päiviö, N.-J. 2012. *Från skattemannarätt till nyttjanderätt – En rättshistorisk studie av utvecklingen av samernas rättigheter från slutet av 1500-talet till 1886 års renbeteslag.* PhD thesis. Uppsala University, Sweden.

Riksrevisionen. 2015. *Granskningar: Gruvafall.* Accessed 18 January 2016. Available at http://www.riksrevisionen.se/sv/GRANSKNINGAR/Planering-och-uppfoljning/Pagaende-granskningar/Pagaende-granskningar/Gruvavfall/.

Rose, D. 1999. Indigenous Ecologies and an ethic of connection, in *Global Ethics and Environment*, edited by N. Low. London, UK: Routledge, 175–187.

Sawhoyamaxa Indigenous Community v Paraguay. 2006. Inter-Am Ct HR (Ser C) No 146 (ruling by the Inter-American Court on Human Rights).

Scheinin, M. 2000. The right to enjoy a distinct culture: Indigenous and competing uses of land, in *The Jurisprudence of Human Rights: A Comparative Interpretive Approach*, edited by T.S. Orlin, A. Rosas, and M. Scheinin. Syracuse, NY: Syracuse University Press, 159–222.

SGU. 2013. *Bergverksstatistik 2012. Statistics of the Swedish Mining Industry 2012.* Periodiska publikationer 2013:2. Sveriges geologiska undersökning. Geological Survey of Sweden.

Statens offentliga utredningar 1986:36, Samernas folkrättsliga ställning. *Delbetänkande av samerättsutredning.*

Tully, J. 1995. *Strange Multiplicity, Constitutionalism in an Age of Diversity.* Cambridge, UK: Cambridge University Press.

Tully, J. 2000. The struggles of indigenous peoples for and of freedom, in *Political Theory and the Rights of Indigenous Peoples*, edited by D. Ivison, P. Patton, and W. Sanders. Cambridge, UK: Cambridge University Press.

UN. 2013. Report of the Special Rapporteur on the rights of indigenous peoples, James Anaya; 'Extractive Industries and indigenous peoples', A/HRC/24/41.

Waldron, J. 1992. Historic injustice: Its remembrance and supersession, in *Justice, Ethics and New Zealand Society*, edited by G. Oddie and R.W. Perrett. Oxford, UK: Oxford University Press.

Weihed, P. and Ahl, P. 2012. Världen behöver mer metall och fler gruvor. *Svenska Dagbladet*, 26 December. Accessed 5 January, 2013. Available at http://www.svd.se/varlden-behover-mer-metall-och-fler-gruvor.

Whitlam, G. 1972. Election Speech. *Australian Federal*, 13 November. Accessed 1 June 2015. Available at http://electionspeeches.moadoph.gov.au/speeches/1972-gough-whitlam.

Wilson, A. and Cervantes, M. 2014. *Fraser Institute Annual Survey of Mining Companies 2013.* Accessed 10 October 2014. Available at https://www.fraserinstitute.org/sites/default/files/mining-survey-2013.pdf.

Part IV

Temporalities of environmental management

11 Challenges in agricultural land management

A Scandinavian perspective on contextual variations and farmers' room to manoeuvre

Elin Slätmo

Introduction

Since the 1950s, the amount of cultivated land on the Scandinavian Peninsula (Sweden and Norway) has decreased; 14 per cent of farmland has been abandoned or developed into other land uses (Li *et al.* 2013). In addition, pasture area in both Sweden and Norway has decreased due to intensification of cultivation and farmland abandonment (Fjellstad *et al.* 2008; Statistics Sweden 2013). A change in agricultural land use to other uses is not in line with national regulations aimed at protecting farmland in both Sweden (Swedish Ministry of the Environment and Energy 2001) and Norway (Norwegian Ministry of Agriculture and Food 1995). From a long-term sustainability and global food supply perspective, it is problematic that farmland is being changed to other land uses. Covering agricultural land with asphalt and developing housing and commercial premises is especially challenging, as it is often irreversible. It cannot be proven with certainty that the current agricultural land in Sweden and Norway will be needed for future production of food, materials and energy, but based on existing levels of knowledge, the precautionary principle should be applied.[1] Three aspects are especially important in this regard. First, the physical conditions for production; agricultural land in Scandinavia is of good quality in terms of soil composition and climate conditions, and with climate change, it may become a more important resource for food, energy and material production relative to other areas in the future. Second, the social and environmental impacts of relocating food production to other places in the world, and of continuing to intensify production using pesticides and fertilisers, require caution about the agricultural land already in use. Third, it is important to manage carefully the values other than food production provided by agricultural land use, for example, food security, recycling of nutrients, rural development, biodiversity and cultural heritage (Deininger *et al.* 2011; Lambin 2012; Meyfroidt *et al.* 2013).

Against this background, this chapter investigates why agricultural land is being changed to other land uses and the potential that exists to influence these changes within today's agricultural policy structures. As the physical and social situation for farming varies with context, exploring differing agricultures is crucial for a deeper understanding of how and why agricultural land use changes

(Lambin *et al.* 2001; Primdahl 2010). Applying the notion of 'driving forces' in order to undertake a contextual analysis, the chapter examines why agricultural land use is changing in Hållnäs in Sweden and Sandnes in Norway. Sandnes and Hållnäs both represent areas where farming is in a difficult situation, but in different ways. Hållnäs is an area of extensive farming in a rural setting with few alternative income opportunities, where preservation of biodiversity and cultural heritage values is the main priority of the authorities. In Sandnes, agriculture is practised in an urban setting, and land claims by other actors are placing limits on continued farming. Drawing on these differing study areas can provide insights into the parallel trends of agricultural land use changes and agricultural land management in Scandinavia.

Driving forces as a notion for analysing changes in agricultural land use

Within landscape research, there has been increasing interest in the concept of 'driving forces' or 'drivers of land use change' since the 1990s (Bürgi *et al.* 2004; Eiter and Potthoff 2007; Schneeberger *et al.* 2007). According to Bürgi *et al.*, driving forces can be defined as 'the forces that cause observed landscape changes, i.e. they are influential processes in the evolutionary trajectory of the landscape' (2004, p. 858). Focussing on why and how human land uses change, using 'driving forces' as a theoretical frame provides potential to analyse landscape changes and identify common land use trends. As the focus of this chapter is on agricultural lands, changes referred to as land use change are from agricultural land to, for instance, forestry, nature conservation areas, fallow land, infrastructure or housing.

The most common way of applying the concept of driving forces is through the use of predefined aspects to categorise the factors and processes that cause land use to change. Previous research on drivers of land use change has mostly been based on the understanding that such change can be explained by correlation and causal relations between different aspects (Hersperger *et al.* 2010).[2] A number of studies have focussed on agricultural land use change, often using a predefined categorisation involving five 'key driving forces', which are formulated as physical environment, socio-economic environment, technology, culture and policy. These categories have been applied in different case studies in Europe (Brandt *et al.* 1999; Bürgi *et al.* 2004; Schneeberger *et al.* 2007; Hersperger and Bürgi 2009), and through the logic of causality, the physical change in agricultural land has been correlated to one of these categories. Farmers as active agents with varying motives influencing their surroundings are seldom included in the studies; instead, what is commonly analysed are the relations between 'the cause' (one of the key driving forces) and 'the effect' (the change of agricultural land to other land uses).

To more fully consider the dynamic relations between farmers' room to manoeuvre and the drivers that cause agricultural land to change, an approach concentrating on factors affecting *farming activities* rather than the *land* should be employed. This approach is based on the perception that land use change is primarily caused by human actions (Moser 1996; Primdahl *et al.* 2004). Land use change from that perspective is perceived as a result of human activities, which are influenced by context-specific social and physical structures (Giddens 1984; Ellegård and Svedin 2012; Shove *et al.* 2012). To understand why land use change occurs, it is therefore necessary to investigate the drivers of human activities, which in the present case represent farmers' activities.

Addressing drivers of human activities is complex. Previous research on farmers' perceptions and attitudes has shown that a number of different motives and approaches are applied to choices of land use and agricultural practices. Research has hence revealed that individual values and motives are crucial for the activities performed in the physical landscape, and that activities and land use cannot be understood through economic models alone (compare Busck 2002; Setten 2004; Wästfelt 2004; Feola and Binder 2010; Hurley and Halfacre 2011; Primdahl and Kristensen 2011). Actors performing agricultural activities are also differently affected by contextual factors. The identity of the farmer is, therefore, important for the type of farming that is performed and what outcomes (if any) contextual factors (such as the key driving forces) have on farming practices.

In order to develop the notion of driving forces to include farmers' agency, driving forces are defined here as the external aspects found in individuals' physical and social context. These structuring contexts affect individuals' activities, in turn affecting the physical environment. This definition means that actors' motives and attitudes (internal driving forces) and structures (external driving forces) are mutually dependent on land use change (compare Hersperger *et al.* 2010). The external driving forces can be physical (geobiophysical preconditions, processes, events and tools) or social (social relations characterised by existing meanings and norms, and the societal organisation of the economy, politics and administration that controls the allocation of resources) and enable, limit or directly trigger the performance of activities (depending on context and internal driving forces). Based on this definition, agricultural land use change is here analysed in light of the dynamic and non-predetermined relations between actors and their surrounding physical and social contexts.

Importantly, the definition of driving forces used in this chapter does relate to the 'key driving forces' categorisation presented above, that is, the physical environment, socio-economic environment, technology, culture and policy. However, analysing the empirical material within these descriptive categories does not directly reveal why land use does or does not change. As noted, the common approach to analysing land use change is to focus on the causal relations between land and these external drivers. However, farmers may 'resist' external drivers and continue to farm the land. The logic of strict causality in land use change studies must therefore be questioned.

In order to more fully consider farmers' room to manoeuvre relative to the external driving forces, I analyse the driving forces through a categorisation based on *how* the social and physical structures influence the agricultural activities rather than the land. This categorisation divides the aspects that farmers in each study area perceive as most important for continued farming into limiting, enabling and direct triggering drivers:

- *Limiting drivers* (compare 'pressures' (Eiter and Potthoff 2007) and Hägerstrand's (1970) three types of restrictions). The contextual aspects categorised as limiting drivers restrict or prevent current agricultural activities.
- *Enabling drivers* (compare 'frictions' and 'repulsions' (Eiter and Potthoff 2007) and 'spaces of possibilities' (Giddens 1976; Westermark 2003; Jones 2009)). The aspects categorised as enabling drivers support current agricultural activities.
- *Direct triggering drivers* (compare 'triggers' (Bürgi *et al.* 2004)). The aspects categorised as direct triggering drivers have the potential to generate direct changes to agricultural land. The difference compared with limiting and enabling drivers is that the effect on land use is more direct in time.

These driver categories are inspired by several scholars (compare Brandt *et al.* 1999; Lambin *et al.* 2001; Bürgi *et al.* 2004; Hersperger *et al.* 2010), with the work of Eiter and Potthoff (2007) the most influential. However, the analytical approach of Eiter and Potthoff does not include human agency, as the categories they use (attraction, pressure, friction, repulsion and working force) indicate that there is a predetermined path of agricultural land use change. In particular, the term 'repulsion' indicates that there are predetermined changes that can be halted, since, according to Eiter and Potthoff, this term means factors that are preventing change (2007, pp. 146, 153). My proposed approach – that is, limiting, enabling and direct triggering drivers – focusses on how structural aspects affect farmers' activities. When human agency is included, there is no continuous predetermined direction of land use change taking place. How land is used is rather determined by continuous interactions between individuals' motives, attitudes and experiences, and the context-specific physical and social structures.

The following section presents the empirical material from Sandnes and Hållnäs. Based on the material, an analysis can then be made of limiting, enabling and direct triggering drivers for agricultural land use change. The results of this analysis are then discussed in relation to farmers' motives for their farming, the theoretical implications of the approach and implications for agricultural land management strategies on the Scandinavian Peninsula.

Agriculture in Hållnäs and Sandnes

Hållnäs: extensive farming in a rural setting

Hållnäs is a peninsula located in the municipality of Tierp, in southcentral Sweden (see Figure 11.1). Hållnäs can be considered a rural area, as it is at least a 45-minute drive to the nearest conurbation with more than 3,000 inhabitants. Hållnäs has a coastal area and a varied inland, largely forested. Long-term extensive farming activities have created meadows and semi-natural pastures that are appreciated for their biodiversity, cultural heritage and recreational values. In 2010, about 10 per cent of the land in Hållnäs was protected within Natura 2000 areas and nature reserves.

Figure 11.1 Locations of Sandnes in Norway and Hållnäs in Sweden.
Map produced by Erik Elldér.

Since the 1950s, there has been a decrease both in population and in active farming. In 1950 there were 2,421 inhabitants living in the parish, while in 2009 Hållnäs had 1,172 permanent residents. Moreover, in 1951 there were 281 farms using 2,356 hectares of agricultural land, compared with 71 farms using 1,514 hectares in 2007 (Statistics Sweden 1956, 1963, 1972, 1983, 2007, 2010). The majority of the population is not active in farming; instead people commute to work outside the peninsula. The farming in the parish is primarily made up of extensive and small-scale livestock farming and/or grass production. The farmers interviewed describe a long agricultural tradition and a Hållnäs characterised by strong cohesion between the inhabitants, with many joint activities and mutual assistance.

Sandnes: agriculture in a urban setting

The city of Sandnes, in southwest Norway, has a population of around 70,000 inhabitants (in 2012; Statistics Norway 2013a; see Figure 11.1). Together with its neighbouring city Stavanger, Sandnes represents a major urban area. Being the hub of the Norwegian oil industry, Sandnes (including Stavanger) experiences fierce pressure on land for urban land use in terms of new housing, infrastructure, public services, business activities and commercial areas (Jøssang *et al.* 2010). Besides the rapidly increasing urban area, the municipality is characterised by hilly areas that are highly appreciated for outdoor recreation and tourist purposes, as well as active farming. The farmers interviewed describe a long and strong tradition of farming in Sandnes. However, the number of active farmers decreased, from 471 to 339, during the period 1996–2012. In 1983, 8,021 hectares were actively farmed, compared to 7,723 hectares of agricultural land in 2012 (Norwegian Agriculture Agency 2013; Statistics Norway 2013b).

Methods

Nine interviews with actors who are both farmers and landowners on the Hållnäs peninsula were conducted from December 2009 to June 2010. These interviewees were all owners of farms with 5–30 hectares of land, and five were hobby farmers, three pensioners and one a full-time farmer. For seven of the interviewees, the land and buildings had been inherited from previous generations.

In Sandnes, 13 farmers were interviewed during April–May 2010. These farmers were the owners of 11 different farms with 20–70 hectares of land. Ten were full-time farmers, two part-time farmers and one a hobby farmer. Twelve of the interviewees had inherited their agricultural land and buildings.

In order to identify the drivers of agricultural land use change in each study area, the discussions during the interviews focussed on their farming activities and the aspects that they perceived as being important for farming in the area historically and for continued farming today and in the future. The interviews were conducted in Swedish and Norwegian, but the quotes used in this chapter are given in English (my translation). In the following section, the findings from the interviews are structured based on the 'key driving forces' of physical environment, socio-economic environment, technology, culture and policy. In order to elucidate why agricultural land use change occurs in Sandnes and Hållnäs in a manner that acknowledges the agency of the farmers, this more descriptive presentation is the basis for the subsequent analysis.

'Key drivers' of agricultural land use change in Hållnäs and Sandnes

Physical environment

On the Hållnäs peninsula, the agricultural land is considered to have low productivity, due to a coastal climate, poor soils and small plots. In Sandnes, farmers reported that both soil and climate conditions are good for food production, although some parts of the municipality are less suited for intensive and conventional agricultural practices because of steep hills and less favourable soil quality.

Socio-economic environment

Income from agriculture was seen as an important factor for continued farming by the interviewees in both Sandnes and Hållnäs. In Hållnäs, no major gains are possible and the economic subsidies for agriculture are described as crucial for farming to break even:

> Today it is impossible to live on this kind of farm ... The knowledge of the local environment, how to perform farming activities, etcetera, you have underneath, and on top of that the reality of the economic conditions and rules exists.
>
> (Farmer 4, Hållnäs)

There is a relative lack of income and service opportunities in Hållnäs, which affects the social environment and the perception of the peninsula as a viable area. The farmers in Hållnäs mentioned the importance of a viable community in encouraging the next generation to take over the farm. However, increased tourism and second home ownership are perceived as processes that can possibly contribute to maintaining the social environment in the parish. Farming was described as being dependent on commuting and employment opportunities outside the peninsula.

In Sandnes, the economic situation was perceived as limiting by the farmers interviewed, despite the relatively high subsidies for farming in Norway. The interviewees attributed this particularly to the rationalisation norm in the agricultural sector, as well as the high costs and salaries outside the sector. Financial investments were perceived as being constantly needed to make production viable. The proximity to the favourable job market in Sandnes (and Stavanger) was also mentioned by the farmers. The contrast in working hours, workload and salary between agriculture and other occupations was described as considerable. This is discouraging intended successors from taking over the farm. At the same time, the access to the job market enabled the farmers to keep their farm as a part-time or leisure activity:

> There's a very strong job market in the municipality, with good wages. You earn much more than in agriculture. When people are established within a job, it is not easy to lose that salary and make the necessary lifestyle changes into farming ... Successors therefore take over the farm, but they do not run it. The farm becomes a nice place to live.
>
> (Farmer 5, Sandnes)

Technology

In Hållnäs, the farmers stated that the combination of physical preconditions and the economic conditions for agriculture today make it unwise to invest in new technology or machinery. In Sandnes, the farmers were concerned about the major economic investments for new machinery and animal houses needed to make agriculture more profitable. The foremost concern was that the large financial risks associated with these investments rest with the individual farmer. Several of the interviewees in Sandnes mentioned that government investment support could be one solution to decrease the risk to the individual and to promote long-term food production. Some of the farmers were concerned about the somewhat one-sided focus on innovation and new technologies in the agricultural sector. Not all farmers believed that it was necessary to have new machinery in order to strengthen food production, and some expressed that this focus carried a risk of not supporting methods and techniques that have been working for decades.

Culture

In both Hållnäs and Sandnes, public appreciation of the values associated with agricultural land emerged as important in motivating farmers to continue farming. The farmers interviewed were clearly aware of the importance of keeping the land actively farmed in a manner that is perceived as attractive for outdoor recreation. The farmers related these values to the willingness among the public to pay for crucial subsidies to agriculture. Furthermore, the long and strong tradition of farming in both areas was described as affecting farming today through a sense of responsibility for past and future generations.

Policy

The farmers in both locations expressed concern about landowners' rights to their land. In Hållnäs, the EU's Common Agriculture Policy (CAP) and concerns relating to nature conservation were highlighted during the interviews. CAP is perceived as problematic in that it is bureaucratically complex and constantly changing and that the overall perspective of the farm is lacking among the relevant local authority divisions. The fact that different authorities and officials are responsible for different parts of the CAP system was also reported as complicating the farmers' relationships with officials:

> Because [the farm] is so small, the EU is crucial. If there were no subsidies, I would not hold on ... But it's hard every year. They're [the subsidies] always changing. The decision comes at the last moment, just days before the forms have to be submitted. Then you sit down and decide what to do in the next five years in a very short time. It's hard, many small farmers have quit due to the EU paperwork ... It's worrying with the subsidies; imagine if you could not cope with all the cross-compliance requirements and were required to repay! It's like a leg trap, you can easily get in but it's hard to get out of it. They have aerial photos of everything.
>
> (Farmer 5, Hållnäs)

As a relatively large proportion of the land in Hållnäs is within nature conservation areas, the interviewees described a duality concerning the natural and cultural values in their lands. Farmers expressed pride in their unique biodiversity and cultural heritage values, but perceived a risk of their land turning into a nature reserve if they practised overly extensive land management:

> I obviously take nature into account, but unfortunately I cannot always take the considerations I want because if I save too many dead trees then someone will come and take away the land from me and make a nature reserve ... We must balance so we are not too adapted to nature. In places where you find high natural values, it is someone who has used the land with respect to nature, why should that land be taken?
>
> (Farmer 2, Hållnäs)

The farmers in Sandnes expressed real concern about their rights to their land in relation to housing developers and officials and politicians involved in spatial planning. The farmers reported that farming in an urban setting is characterised by conflicts with other actors and their land claims.

The prices for agricultural land sold for housing and other commercial uses are considerably higher than those for land sold for agriculture:

> There's no-one who wants to sell land to farming if you can sell to construction! ... Many of the neighbours are no longer farmers as their land has been categorised as future development area [in local spatial planning]. There are also some who can sell if they want to ... and there are many who lease out their land. Every time there is a new manager in the development firm they call ... and say: 'if you want to sell, then just call'.
>
> (Farmer 3, Sandnes)

Categorising drivers based on how they influence land use activities

The 'key driving forces' categories for the farmers' perceptions of agricultural land use change in Hållnäs and Sandnes presented above illustrate the range of factors influencing change in agricultural land use. In order to develop the notion of driving forces in a way that more clearly includes human agency, conveying that land use changes are not necessarily predetermined, the analytical categories of limiting, enabling and direct triggering drivers were next applied to analyse why agriculture land use change occurs in Hållnäs and Sandnes.

Limiting drivers for agricultural activities

Table 11.1 shows the physical and social structural aspects that are limiting for continued farming according to the farmers interviewed.

Table 11.1 Limiting drivers for agricultural activities in Sandnes and Hållnäs.

Sandnes	Hållnäs
Economic situation for agricultural production	Economic situation for agricultural production
Lack of successors to take over the farms	Lack of successors to take over the farms
Rationalisation policy for farming	
Demand for land by other actors	
	Physical conditions for agriculture
	Relative lack of job opportunities and services nearby
	Nature conservation measures
	A policy system that is micromanaged and difficult to comprehend

The economic situation for agriculture represents a struggle in both places. A thought-provoking finding is that the differences between the study areas are strongly connected to aspects outside the agricultural sector, for instance, lack of job opportunities and nature conservation measures in Hållnäs and other actors' land claims in Sandnes.

Enabling drivers for agricultural activities

Table 11.2 shows the physical and social structural aspects that are enabling continued farming according to the farmers interviewed.

A sense of responsibility, economic subsidies and public appreciation of recreational values in farmland are working as enabling drivers in both places. However, Table 11.2 also shows differences between Hållnäs and Sandnes, which are primarily related to the social environment. In Sandnes, the proximity to an urban area is perceived as strongly supporting the farmers' social life, as it is a convenient distance to alternative jobs, services and/or leisure activities. The social environment in Hållnäs is not perceived as vibrant by the farmers, yet in

Table 11.2 Enabling drivers for farming in Sandnes and Hållnäs.

Sandnes	Hållnäs
Sense of responsibility to farm the land	Sense of responsibility to farm the land
Economic subsidies for agriculture	Economic subsidies for agriculture
Public appreciation of environmental and recreational values in farmland	Public appreciation of environmental and recreational values in farmland
	Campaigns for rural community development
	Tourism and second home owners
Legislation for protecting agricultural land	
A pleasant social environment close to an urban centre	

recent years Hållnäs has increased in popularity as a tourist destination which, combined with more second home ownership and related new income opportunities, are perceived as aspects that can maintain the social environment in the parish. As mentioned above, both Sweden and Norway have legislations aimed at protecting farmland. However, legal instruments were foremost mentioned as an enabling factor in the interviews with farmers in Sandnes.

Direct triggering drivers for agricultural land use change

Table 11.3 shows the physical and social structural aspects that are direct triggering drivers for agricultural land use change according to the farmers.

In Hållnäs, decisions on land use in spatial planning and changes in subsidies for agriculture were described as directly triggering changes in land use. For example, changes within the CAP in 2005 to an areal payment scheme instead of a production-based scheme led to increased use of former fallow land. In peri-urban Sandnes, the primary determinants of how the land is used, whether for farming or not, are decisions within spatial planning. The demand for land from other actors is great, and spatial planning decisions directly trigger agricultural land to change to other uses.

Table 11.3 Direct triggering drivers for agricultural land use change in Sandnes and Hållnäs.

Sandnes	Hållnäs
Decisions on land use in spatial planning	Decisions on land use in spatial planning
	Changes in subsidies for agriculture

The motives behind farming in Hållnäs and Sandnes

The present study of farmers in Sandnes and Hållnäs demonstrates that the situation for farming clearly differs between the two areas, but what they have in common is the troubled situation for continued farming and what the farmers interviewed saw as harsh conditions for farming in the future. This led me to speculate not only about the causes of agricultural land use change but also about the resistance shown by the farmers in Sandnes and Hållnäs to drivers of change. Listening to the farmers and including the enabling drivers in the analysis helped identify the possibilities that exist to promote future farming. Hence, to further deepen an understanding of farmers' resistance to the limiting external driving forces, the motives behind farming in Sandnes and Hållnäs need to be discussed.

In Hållnäs, all farmers interviewed shared the characteristic that their farming mainly stemmed from non-financial motives – an interest in working with soil and animals, a curiosity to try different land management methods and/or a sense of responsibility to keep the land farmed. This is possibly due to the fact that only one of the farmers interviewed in this area was a full-time farmer. However, even the full-time farmer referred to non-financial motives for farming, as his income primarily stemmed from forestry and what would be classified as 'other businesses' in the statistics (selling firewood and renting out machinery). One of the farmers said that his ambition was never to live off farming, but that it was a hobby:

> Have never had the ambition to want more, this is more like a hobby. Not everyone can live like this ... that people want to live like this is a crucial factor. It is quiet here, and some people are tired of big city life ... Then there are people like me who do not know much else, who have always been here and still thrive.
>
> (Farmer 7, Hållnäs)

The farmers in Sandnes also had different motives and ambitions for their farming. The full-time farmers tended to be more concerned about the economic situation for agriculture, while the part-time and hobby farmers were more focussed on the availability of other jobs and services. When asked why they keep on farming, all the interviewees referred to a genuine interest in farming as their primary motive. Note, however, that the farmers interviewed in Sandnes were landowners who wanted to continue as farmers and not sell their land for development purposes. Several of the farmers stated that they are driven by idealism, as they perceive the preservation of agricultural land to be more important than the money they could obtain from selling their land:

> In peri-urban areas the willingness to pay is so great that most [farmers] would sell [their land]. But there are still a few in Sandnes who do not want to sell despite large amounts [of money], and that is due to idealism. It's about the desire to continue farming the land and pass on a legacy and tradition.
>
> (Farmer 4, Sandnes)

In both Hållnäs and Sandnes, some similarities in how the farmers talked about their farming can be highlighted. A real interest in cultivating the land and rearing livestock emerged as the primary motive for the interviewees' activities. A desire to continue farming the land was also described by the interviewees as being closely linked to a sense of responsibility to both past and future generations to keep the land farmed. Based on this, the farmers are trying to create as favourable an economic situation as possible.

The meaning of contextual variations and farmers' room to manoeuvre

The suggested approach to analysing driving forces is a way to categorise the complexity and dynamics of agricultural land use change. The categories of limiting, enabling and direct triggering drivers make it possible to highlight aspects and processes influencing farmers' decisions, which in turn improves understanding of why agricultural land use passes to other land uses (or not), without necessarily reducing the complexity of the changes that are taking place. The suggested approach hence allows for complexity, as there are neither predetermined categories for the driving forces nor any predetermined direction of land use development. Thus, the approach is a response to the call to make studies of driving forces more temporally and spatially specific (Lambin *et al.* 2001; Bürgi *et al.* 2004; Primdahl 2010). The categories of enabling, limiting and direct triggering drivers can be used to show that the situation for farming differs with context. This has two main advantages. First, the studies in Hållnäs and Sandnes show that agriculture is affected by, and dependent on, a variety of activities and processes beyond the agricultural sector, for instance, nature conservation measures, tourism and housing development. Investigating the factors influencing the agricultural activities, and not only the agricultural land, facilitates integration of drivers that are not part of or directly connected to the agricultural sector. Second, categorising the driving forces according to how the factors affect farming activities allows for a discussion around the contextual dependency of agricultural land use change. A particular aspect can differ in influence depending on the context under study (compare Lambin *et al.* 2006). For example, in the peri-urban area of Sandnes, where residential and commercial development is prioritised over preservation of agricultural land in the spatial planning system, my approach would categorise this as limiting for agriculture, as planning decisions are conflicting with agricultural land use. However, if the same situation arose in Hållnäs, my approach would instead categorise it as enabling, as it could mean better opportunities for alternative income for farmers. Thus, how the relations should be categorised, as enabling or limiting, depends on the context of the study.

By implication, it is important to highlight that there are no clear boundaries between the categories. As some aspects can be both enabling and limiting for farming in the specific case, they should instead be viewed as frames aiding thinking in new ways. For example, in Hållnäs, biodiversity and cultural heritage values in agricultural land can be perceived as both enabling and limiting for farming. The farmers expressed pride in the unique biodiversity and cultural heritage values, but also saw a risk of their land being turned into conservation areas if they manage it in line with the official view of what is most valuable. In Sandnes, the proximity to urban centres is both enabling and limiting for continued farming. It is viewed as worrying by the farmers as it creates claims on their land by other actors. At the same time, it provides access to jobs, services and recreational areas which enable farmers to keep their farm as a part-time or leisure activity and is perceived as positive for farmers' social life.

It is crucial to acknowledge and support the differing motives and attitudes for agriculture in order to promote continued farming, as committed farmers are crucial for keeping agricultural land in active use. However, a positive attitude and an interest in farming alone are not sufficient to preserve agricultural land. Other aspects and processes external to the individual farmer are important in determining why agricultural land is lost to other land uses. Analysing changes in agricultural land use in Hållnäs and Sandnes revealed that decisions by actors and institutions other than the individual landowner or farmer are crucial for land use change. However, these actors and institutions are not the same in the different places.

In Hållnäs, the EU and the international food sector, combined with small farm size and relatively poor soil quality, make it impossible to develop economically viable farms. It is leisure farmers who can possibly keep the agricultural landscape as it looks today, with the help of economic support through subsidies. This analysis has shown that connections to other activities, beyond agriculture, such as possibilities to commute to work, a maintained service level and tourism development, are enabling for future farming. It is also important to highlight that the dominant management strategy, that is, protected areas to conserve cultural heritage and biodiversity values, can limit continued farming. It is important to state that maintenance of farming is dependent on income opportunities but that conservation areas are primarily based on preservation logics. Therefore, they seldom allow or support development of new income activities (see also Mittenzwei *et al.* (2010) for similar findings on cultural landscape values in Norway).

The situation in Sandnes implies that political decisions in spatial planning are crucial for preserving agricultural land, which is a basic prerequisite for continued farming. In particular, spatial planning at the municipal level, which is where many land use decisions in Norway are made, is an important tool to steer the use of land in Sandnes. Due to high land prices, farmers reported that it is economically irresponsible not to sell their land if it is categorised as a development area in municipal plans. However, the farmers interviewed in Sandnes were resisting the high prices for their land, as they perceived other values as more important.

Concluding remarks

In this chapter, I have analysed agricultural land use changes in Hållnäs in Sweden and Sandnes in Norway by applying the notion of driving forces. The results show that in order to obtain basic knowledge of why agricultural land use change occurs, there is a need to clearly identify the physical and social contexts for agriculture. To further facilitate integration of context in studies of the driving forces of agricultural land use change, the focus should be on factors important for continued farming, that is, what affects agricultural activities and not what affects agricultural land. Categorising the drivers into limiting, enabling and directly triggering aspects for farming helps identify what is limiting for further agricultural activities and what is enabling for continued farming. Applying these general categories could be a starting point for discussing the complex realities on the ground. The approach permits analysis of agricultural land use change from the farmers' reality, thereby identifying temporally and spatially specific aspects causing agricultural land use to change or to be maintained as farmland. It facilitates problematisation of the temporal dimension of agricultural land use; the long and strong tradition of farming in Hållnäs and Sandnes is important for current and future farming, as the desire to keep the land in active use stems from a real interest in working the land, working with animals and maintaining the historical legacy of keeping the land actively farmed. A feeling of responsibility to keep the land farmed for both past and future generations enables the current farming. It can however be discussed whether that feeling is enough to keep the land farmed in the long term, especially when the hardship involved in living solely off farming and the lack of interested successors is considered. Studying agricultural land in Sandnes and Hållnäs has also shown the importance of placing agriculture in relation to the greater society and analysing drivers of land use changes beyond the agricultural sector. In both Sandnes and Hållnäs, actors and institutions other than the farmers and landowners had predominant positions in decisions over agricultural land; in Sandnes, agricultural land as a resource for future production was mainly in conflict with urban land use interests, while in Hållnäs, biodiversity and cultural heritage protection interests were predominant.

In order to influence agricultural land use change through public management, further attention needs to be given to conflicting land use interests and further developing methods for democratically solving these conflicts. Furthermore, if agricultural land use is to be maintained over the long term as legislated in both Norway and Sweden, the studies in Sandnes and Hållnäs indicate a need for discussions of how to combine management strategies for preservation and development, for example, what should be preserved and what can be further developed to retain farming.

Notes

1 The precautionary principle is a model of public decision making in situations where potential environmental or health risks are considered proven but when scientific knowledge is insufficient for a firm conclusion regarding the magnitude of the risk to be drawn. Essential for the precautionary principle is that lack of knowledge should not be used as a reason to refrain from safeguarding measures, but instead care should be exercised to the greatest extent possible in order to mitigate or prevent future environmental and health problems. The precautionary principle or approach has been established internationally, both within the UN and the EU (COMEST 2005; Handl 2012). In Sweden, the precautionary principle is expressed in the Swedish Environmental Code. In Norway, the Norwegian Nature Conservation legislation is based on the principle.
2 See Hersperger *et al.* (2010) for a discussion on the different approaches to driving forces of land use change.

References

Brandt, J., Primdahl, J. and Reenberg, A. 1999. Rural land-use and landscape dynamics – analysis of 'driving forces' in space and time, in *Land-Use Changes in Europe and Their Environmental Impact in Rural Areas in Europe*, edited by R. Krönert, J. Baudry, I.R. Bowler and A. Reenberg. Paris and New York: UNESCO and Panthenon, 81–102.

Bürgi, M., Hersperger, A.M. and Schneeberger, N. 2004. Driving forces of landscape change — current and new directions. *Landscape Ecology*, 19(8), 857–868.

Busck, A.G. 2002. Farmers' landscape decisions: Relationships between farmers' values and landscape practices. *Sociologia Ruralis*, 42(3), 233–249.

COMEST. 2005. *The Precautionary Principle.* World Commission on the Ethics of Scientific Knowledge and Technology (COMEST). France: United Nations Educational, Scientific and Cultural Organization.

Deininger, K., Byerlee, D., Lindsay, J., Norton, A., Selod, H. and Stickler, M. 2011. *Rising Global Interest in Farmland: Can It Yield Sustainable and Equitable Benefits?* Washington, DC: The International Bank for Reconstruction and Development/The World Bank.

Eiter, S. and Potthoff, K. 2007. Improving the factual knowledge of landscapes: Following up the European Landscape Convention with a comparative historical analysis of forces of landscape change in the Sjodalen and Stølsheimen mountain areas, Norway. *Norsk Geografisk Tidsskrift-Norwegian Journal of Geography*, 61(4), 145–156.

Ellegård, K. and Svedin, U. 2012. Torsten Hägerstrand's time-geography as the cradle of the activity approach in transport geography. *Journal of Transport Geography*, 23, 17–25.

Feola, G. and Binder, C.R. 2010. Towards an improved understanding of farmers' behaviour: The integrative agent-centred (IAC) framework. *Ecological Economics*, 69(12), 2323–2333.

Fjellstad, W., Norderhaug, A. and Ødegaard, F. 2008. *Tidligere og Nåværende Jordbrukareal – Miljøforhold og Påvirkninger på Rødlistearter.* Trondheim, Norway: Artsdatabanken.

Giddens, A. 1976. *New Rules of Sociological Method: A Positive Critique of Interpretative Sociologies.* London, England: Hutchinson.

Giddens, A. 1984. *The Constitution of Society: Outline of the Theory of Structuration.* Cambridge, UK: Polity.

Hägerstrand, T. 1970. What about people in regional science? *Papers in Regional Science*, 24(1), 724.

Handl, G. 2012. *Declaration of the United Nations Conference on the Human Environment (Stockholm Declaration), 1972 and the Rio Declaration on Environment and Development, 1992.* United Nations Audiovisual Library of International Law.

Hersperger, A.M. and Bürgi, M. 2009. Going beyond landscape change description: Quantifying the importance of driving forces of landscape change in a Central Europe case study. *Land Use Policy*, 26(3), 640–648.

Hersperger, A.M., Gennaio, M., Verburg, P.H. and Bürgi, M. 2010. Linking land change with driving forces and actors: Four conceptual models. *Ecology and Society*, 15(4), 1–17.

Hurley, P.T. and Halfacre, A.C. 2011. Dodging alligators, rattlesnakes, and backyard docks: A political ecology of sweetgrass basket-making and conservation in the South Carolina Lowcountry, USA. *GeoJournal*, 76(4), 383–399.

Jones, M. 2009. Phase space: Geography, relational thinking, and beyond. *Progress in Human Geography*, 33(4), 487–506.

Jøssang, L.G., Lindanger, B., Nerheim, G. and Tysdal, O. 2010. *Sandneshistorien - til stor by.* Bergen, Norway: Fagbokforlaget.

Lambin, E.F. 2012. Global land availability: Malthus versus Ricardo. *Global Food Security*, 1(2), 83–87.

Lambin, E.F, Geist, H.J. and Rindfuss, R.R. 2006. Introduction: Local processes with global impacts, in *Land-Use and Land-Cover Change: Local Processes and Global Impacts,* edited by E. Lambin and H. Geist. Berlin, Germany: Springer Verlag, 1–8.

Lambin, E.F., Turner, B.L., Geist, H.J., Agbola, S.B., Angelsen, A., Bruce, J.W., *et al.* 2001. The causes of land-use and land-cover change: Moving beyond the myths. *Global Environmental Change*, 11(4), 261–269.

Li, B.-B., Jansson, U., Ye, Y. and Widgren, M. 2013. The spatial and temporal change of cropland in the Scandinavian Peninsula during 1875–1999. *Regional Environmental Change*, 13(6), 1–12.

Meyfroidt, P., Lambin, E.F., Erb, K.-H. and Hertel, T.W. 2013. Globalization of land use: Distant drivers of land change and geographic displacement of land use. *Current Opinion in Environmental Sustainability*, 5(5), 438–444.

Mittenzwei, K., Lien, G., Fjellstad, W., Øvren, E. and Dramstad, W. 2010. Effects of landscape protection on farm management and farmers' income in Norway. *Journal of Environmental Management*, 91(4), 861–868.

Moser, S.C. 1996. A partial instructional module on global and regional land use/cover change: Assessing the data and searching for general relationships. *GeoJournal*, 39(3), 241–283.

Norwegian Agriculture Agency. 2013. *Statistikk fra Søknader om Produksjonstilskudd i Jordbruket*. Oslo.

Norwegian Ministry of Agriculture and Food. 1995. *The Land Act. Act No. 23 of 12 May 1995 Relating to Land. Chapter IV. Protection of Cultivated and Cultivable Land, etc.* Oslo.

Primdahl, J. 2010. Globalisation and the local agricultural landscape: Current change patterns and public policy interventions, in *Globalisation and Agricultural Landscapes: Change Patterns and Policy Trends in Developed Countries*, edited by J. Primdahl and S. Swaffield. Cambridge and New York: Cambridge University Press, 149–167.

Primdahl, J., Busck, A.G. and Kristensen, L.S. 2004. Landscape management decisions and public-policy interventions, in *The New Dimensions of the European Landscapes*, edited by R.H. Jongman. The Netherlands: Springer, 103–120.

Primdahl, J. and Kristensen, L.S. 2011. The farmer as a landscape manager: Management roles and change patterns in a Danish region. *Danish Journal of Geography*, 111(2), 107–116.

Schneeberger, N., Bürgi, M., Hersperger, A.M. and Ewald, K.C. 2007. Driving forces and rates of landscape change as a promising combination for landscape change research – An application on the northern fringe of the Swiss Alps. *Land Use Policy*, 24(2), 349–361.

Setten, G. 2004. The habitus, the rule and the moral landscape. *Cultural Geographies*, 11(4), 389–415.

Shove, E., Pantzar, M. and Watson, M. 2012. *The Dynamics of Social Practice: Everyday Life and How It Changes*. Los Angeles, CA: Sage.

Statistics Norway. 2013a. *Befolkningsendringer i Kommunene*. Oslo.

Statistics Norway. 2013b. *Strukturen i Jordbruket. Jordbruksareal, Otter Bruken. 1979, 1989, 1999, 2008–2012*. Oslo.

Statistics Sweden. 1956. *Jordbruksräkningen 1951*. Jordbruk med binäringar, Stockholm.

Statistics Sweden. 1963. *Jordbruksräkningen 1961*. Jordbruk med binäringar, Stockholm.

Statistics Sweden. 1972. *Lantbruksräkningen 1971*. Sveriges Officiella Statistik, Stockholm.

Statistics Sweden. 1983. *Lantbruksräkningen 1981*. Sveriges Officiella Statistik, Stockholm.

Statistics Sweden. 2007. *Jordbruksföretag och Företagare*. Sveriges Officiella Statistik, Stockholm.

Statistics Sweden. 2010. [Statistics received from the Agricultural Departement at Statistics Sweden. 16/03/2010 & 17/03/2010].

Statistics Sweden. 2013. *Markanvändningen i Sverige*. Sveriges Officiella Statistik, Stockholm.

Swedish Ministry of the Environment and Energy. 2001. *Ds 2000:61 The Swedish Environmental Code, Chapter 3. Basic provisions concerning the management of land and water areas, section 4*.

Westermark, Å. 2003. *Informal Livelihoods; Women's Biographies and Reflections about Everyday Life, a Time-Geographic Analysis in Urban Colombia*. Göteborg University: the Department of Human and Economic Geography, B 102.

Wästfelt, A. 2004. *Continuous Landscapes in Finite Space: Making Sense of Satellite Images in Social Science*. PhD thesis, Stockholm University, Stockholm, Hugo förlag.

12 Performing natures

Adaptive management practice in the 'eternally unfolding present'

Ruth Beilin and Simon West

Introduction

Modernist approaches to environmental management have been comprehensively critiqued in recent decades across multiple disciplines. Pretensions to complete (expert) knowledge and control have become widely regarded as 'pathologies' (Holling and Meffe 1996) or 'fallacies' (Stirling 2014). Environmental management scholars have sought alternative ways to structure knowledge–action relationships that are more cognisant of the complex, dynamic and, to some extent, unknowable relations between humans, non-humans, and landscapes (e.g. Berkes *et al.* 2003). Adaptive management, primarily developed within resilience science, has become one such increasingly influential alternative, embraced in Australian and Scandinavian environmental management, policy and planning (Allen and Garmestani 2015). Adaptive management is based on the idea that, in the context of unpredictable change, knowledge will be incomplete and provisional and management experience should be used to produce scientific knowledge and to shape subsequent management actions (Holling 1978; Walters 1986). Consequently, adaptive management attempts to enhance learning from experience by framing management plans as scientific hypotheses and management actions as experiments in processes of learning-by-doing. However, while adaptive management is intuitively attractive as a way of bridging science and practice in order to 'allow management action to continue' in complex and uncertain situations, it has proven difficult to connect adaptive management decisions to desired outcomes in the landscape, and a gap has emerged between theory and practice (Allen and Garmestani 2015, p. 3).

We argue that the gap between the theory and practice of adaptive management has emerged in part because the treatment of management experience as a basis for knowledge and action is problematic when conceived within epistemologies of science. Scientific approaches to practice assume that reality is made up of discrete entities with pre-given properties, that the most basic way to 'know' this reality is through the subject–object relation, and that practitioners undertake action on the basis of cognitive representations of these discrete entities (Sandberg and Tsoukas 2011; Cook and Wagenaar 2012). For adaptive management, it is consequently assumed that the utility of manager experience is representative – to improve scientific models of the problem at hand – and that the resultant representations will in turn improve practice by identifying discrete variables for managers to manipulate. However, accounts of practitioner life suggest that managers do not experience practice merely as a set of discrete variables, but as an inherently situated and meaningful 'unfolding totality' (Sandberg and Tsoukas 2011, p. 341). Moreover, managers tend not to experience practice as an atemporal space, but as a 'continuous, ongoing stream of activity' (Cook and Wagenaar 2012, p. 14). Yet scientific epistemologies abstract away the temporal flows (Sandberg and Tsoukas 2011, p. 342) of adaptive management practice, including the timescales of scientific knowledge generation, action within environmental management agencies and ecosystem and landscape responses. Indeed, the focus in adaptive management on 'getting the facts right' (or, as near to 'right' as possible; Cote and Nightingale 2012) has made it difficult for the literature to account for the ways in which the technical application of adaptive management tools – such as monitoring, modelling and experimentation – become entangled in, for example, competing ideal natures for a given site, interpersonal and organisational human relations and uncertainties about the relative sensitivity and efficacy of the tools themselves. The messiness of practice consequently appears through scientific epistemologies as a 'social barrier' to adaptive management implementation rather than the context through which adaptive management is constituted in any particular locality (Johnson 1999).

We suggest that shifting registers to an epistemology of practice may better describe enactments of adaptive management in 'the real world' and therefore reduce the gap between theoretical prescriptions for, and practical performances of, adaptive management. Cook and Wagenaar (2012) clarify that an epistemology of practice explains knowledge and context in terms of practice, rather than the other way around. Whereas epistemologies of science tend to view practice as a 'mere conduit for knowledge *of* the world', epistemologies of practice view practice as 'the vital element of *knowing in* the world' (Cook and Wagenaar 2012, p. 16; emphasis in the original). Here, natures are not simply represented by experience but performed through experience. Performativity emphasises that experience of the environment is a product of our engagement with the world, that we experience the constraints and affordances of the environment in emergent time, and that human and material elements interpenetrate and affect the ways we know and understand the world (Griggs *et al.* 2014). The focus shifts from 'getting the facts right' to the moment-to-moment practices of negotiation through which practitioners 'get the right facts' (West 2015).

In this chapter we trace the epistemological origins of adaptive management through ecological pragmatism, dual control theory and resilience science. We then critique the treatment of management experience in the adaptive management literature to date and identify a number of ways in which epistemologies of practice may help draw out the pragmatist heritage of adaptive management and reduce the gap between theory and practice. We explore the practical import of this theoretical work in two case narratives of Australian fire management. In the first narrative, we use a performative lens to reveal the ways in which scientific planned burning in the state of Victoria disguises the moment-to-moment negotiation between social-ecological imaginaries of fire, historical ecological data and political imperatives. In the second, we explore how indigenous and scientific co-management of an Indigenous Protected Area (IPA) in the Tanami Desert is working to effect a culturing of adaptive management, where such moment-to-moment negotiation is embraced and space is made for place-specific practices to evolve 'on country'. The chapter forms a starting point to develop performative case studies of adaptive management 'in action' (e.g. Scoones 1999; West 2015).

First, we clarify our use of terms. Fire management occurs in the landscape, understood as 'a place of human habitation and environmental interaction' (Olwig 1996, p. 630). The evocation of landscape across disciplines reflects that 'the major grand challenges facing our society are embedded in landscape' and provides an integrated research platform to study climate change, energy needs, health and safety, and other intersecting fields (European Science Foundation 2010, p. 41). Beilin and Bohnet (2015) describe landscape research as creating a platform for the social-ecological imaginaries that inform us about place and human and non-human agency. In this chapter, landscape is the place wherein entanglement of the material and embodied, dynamic and performative aspects of fire occur. Adaptive management, conceived by ecologists, predominately imagines landscape as a template in which, regardless of distance (scale), the area of interest is large enough to ascertain spatial patterns that influence ecological processes (Wiens and Milne 1989). Planned and ecological burning at a simple level is based on recognising these landscape scale patterns or attempting to create them. It becomes the task of adaptive management practitioners to achieve these larger-scale management outcomes. In resilience science, an ecological system is imagined to exist within a broader, coupled social-ecological system, wherein everything is connected (Berkes *et al.* 2003). As such, the substantive meaning of landscape (compare Olwig 1996) evolves in this chapter to embody the material and non-material ways in which practitioners construct fire management as part of biophysical and cultural place.

From representing to performing natures in adaptive management

Adaptive management emerged from the convergence of dual control theory and resilience science in the 1970s (Curtin and Parker 2014). Dual control theory is based on the premise that, in managing a system with unknown characteristics, a notional 'controller' (i.e. an environmental manager) has two primary options: (a) acting to control the system as well as possible given existing knowledge, and (b) investigating or actively probing the system to gain more knowledge about it (Feldbaum 1961; Walters and Hilborn 1978). The dilemma posed by dual control theory is that probing the system to learn about it causes perturbations, reducing the ability to control the system. Therefore, dual control theory confronts the 'controller' with a basic decision – act immediately on the basis of existing knowledge, or wait for more information in the hope of making a more informed choice. Scholars developing dual control theory in an environmental management context have consequently attempted to develop means of 'optimizing' this basic choice in relation to the 'controls' (management actions) available and the management objectives of particular ecological systems (Williams and Johnson 2015). These optimisation techniques require assessment of the potential value of acting in the short term against the potential value of long-term learning (Moore *et al.* 2008).

The development of dual control theory dovetailed with early articulations of ecological resilience, first defined by ecologist C.S. Holling (1973, p. 14) as the 'measure of the persistence of systems and of their ability to absorb change and disturbance and still maintain the same relationships between populations or state variables'. Resilience challenged the primacy of stability as the central organising principle in ecological systems, instead emphasising non-linear, dynamic, cross-scale change between fluctuating stable states (Folke 2006). Management through a resilience lens begins with the assumption of incomplete, provisional knowledge and encourages broad participation and multiple perspectives, including indigenous knowledge, in building models and devising monitoring regimes (Walker and Salt 2006). These processes of 'social learning' effectively aim to enhance the learning choice of dual control theory. Adaptive management thus enrols the basic problem formulation and mathematical modelling methods of dual control theory with an 'experimental approach to social reform', as a way to 'operationalize the principles of resilience science through cross-scale and place-based experimentation and institutional learning' (Curtin and Parker 2014, p. 917).

However, practitioners have struggled to enact these practices within extant forms of environmental management (Allan and Curtis 2005; Jacobson *et al.* 2006). Reported problems include disparities between the temporalities of scientific knowledge production and the temporalities of action in environmental agencies; risk aversion and procrastination in publicly accountable agencies; a lack of resources to embark on adequate monitoring or experimentation; and a lack of stakeholder engagement leading to conflict between competing interests (Allen and Gunderson 2011). The difficulties of 'making adaptive management practices tractable in diverse ecological, cultural and historical contexts' have been characterised in terms of 'social barriers', 'pathologies' and 'gaps' – but they also emerge from unresolved epistemological and ontological tensions that sit at the heart of adaptive management (West 2015, p. 7). These unresolved tensions are reflected in the disaggregation of adaptive management into a number of ideal types, each addressing different aspects of the temporal flow of knowledge and action, including evolutionary and strategic adaptive management (Walters and Holling 1990), passive and active (Williams 2011), decision-theoretic and resilience-based (Johnson *et al.* 2013). The ontologies that sit behind these ideal types, ranging from Walters' (1986, p. 3) scepticism about the existence of 'a single, structurally stable system out there in nature to be understood', to Allen and Garmestani's (2015) implicit commitment to a complex but apparently timeless 'true state of nature', are often masked, even while shaping adaptive management practices and expectations.

These epistemological and ontological tensions have emerged because of a desire to bridge the gap between applied and basic scientific research, on the one hand, and between scientific research and management practice, on the other (Holling and Sundstrom 2015). However, because learning-by-doing has been operationalised through the vehicles of mathematical modelling, scientific monitoring and experimentation, the adaptive management literature to date has addressed relationships between theory and practice primarily through the representational idiom, or what Sandberg and Tsoukas (2011) call 'scientific rationality'. Here, the primary task of the researcher is imagined to consist of observing and theoretically representing the world from the 'outside'. Sandberg and Tsoukas summarise scientific rationality in three premises: first, reality is made of discrete entities with particular pre-given properties; second, the most basic form of knowing about this reality is through the 'subject–object relation' – that humans, as subjects, exist independently of the objects they seek to represent; third, this subject–object relationship is the logic underlying practice – that 'practitioners face a world of discrete objects whose pre-given features they represent through cognitive activity and, on the basis of those representations, undertake action' (2011, p. 340). While scientific rationality assumes that the basic cognitive processes of practitioners and scientists are the same – that they both deal with 'variables' – practitioner knowledge is imagined to be more subjective and biased because practitioners are so 'close' to practice. Therefore the role of the scientific researcher, operating at a distance from practice, is to provide practitioners with more objective, abstract knowledge.

There are several ways in which the heritage and the promise of adaptive management, as outlined by Walters (1986) in particular, challenge the scientific rationality outlined above. For instance, Holling (1978) and Walters (1986) explain that adaptive management is valuable because knowledge will always be incomplete, uncertain and provisional. They explicitly frame the practice of adaptive management in terms of generating knowledge through active experience and experimentation – an approach traced by Curtin and Parker (2014) to the pragmatist philosophy of John Dewey – and emphasise that the value of the model-building process is not the final representation but the process itself which encourages close reflection on the particular management context. Walters asserts that the models will always be wrong, embraces management experience as a valid means of generating knowledge and consequently emphasises the temporal complexities of knowing and acting; this approach goes some way to challenging the subject–object relation of scientific rationality. However, while Walters highlights the importance of learning-by-*doing*, he never doubts that the primary character, purpose and utility of this experience is representational. Thus, experience in adaptive management is ultimately filtered through the subject–object gaze, and learning is limited to producing a better mathematical model. The assumption implicit within scientific rationality that practitioners carry a more subjective form of knowledge than scientists is embraced by adaptive management as a useful means of learning about a system that, while complex, constitutes a potentially unifiable whole. This is evident in collaborative adaptive management, where practitioners' subjective 'mental models' of landscape dynamics are collected and then combined to produce an apparently more accurate representation of the problem at hand (Etienne *et al.* 2011). The implication of an atemporal nature that can be disaggregated into discrete variables, and the assumption that this process will improve management practice, remains.

Practice-based approaches highlight several problems with conceiving of adaptive management practice through the lens of scientific rationality. First, scientific rationality underestimates the 'meaningful totality in which practitioners are immersed' (Sandberg and Tsoukas 2011, p. 341). Despite the call by Scoones (1999) for anthropological accounts of adaptive management that retain contextual and experiential richness, accounts of adaptive management practice tend to leave out 'the massive, largely tacit body of understandings, memories, expectations, and routines that, together, inform the essential context of our actions' (Cook and Wagenaar 2012, p. 14). Second, scientific rationality ignores the 'situational uniqueness' in practitioners' performance of tasks (Sandberg and Tsoukas 2011, p. 341). The importance of situational uniqueness to adaptive management becomes apparent in studies of applied field ecology (e.g. Robertson 2006), where managers and scientists face a range of 'dilemmas and dislocations' in getting supposedly universal scientific tools to work in the field (Griggs *et al.* 2014, p. 11). Accordingly, Jacobson *et al.* (2009, p. 485) argue that 'improvement in the practice of adaptive management requires greater attention to uncertainties as they relate to practice context'. Finally, by 'conceptualizing practice as an atemporal space, scientific rationality abstracts away from the *temporal flow* of practice' (Sandberg and Tsoukas 2011, p. 342). However, the tempo and directionality of practice – in what Cook and Wagenaar (2012, p. 21) refer to as the 'eternally unfolding present' – are critical aspects of how practitioners make sense of their activity and enact adaptive management in dynamic contexts. While temporal emergence is a crucial facet of the complexity perspectives underlying adaptive management and is found in Walters' (1986) emphasis on timing, the significance of time disappears in most accounts of adaptive management practice. Recognising the importance of meaning, context and temporality to adaptive management practice requires switching from representational to performative idioms.

While we accept that representation plays a very important role in practice, we argue here that it is not the only – and not the *most* important – way in which practitioners act, know and respond to the situations in which they find themselves (Wagenaar 2011). It is important to draw out alternative approaches to practice, because a common response among scientists to the 'implementation problems' of adaptive management has been to restrict its application to a small subset of management problems with 'few interacting variables' and few or no competing human interests (e.g. Gregory *et al.* 2006). In other words, it has been restricted to landscapes that are more easily represented through mathematical models. In our view, this represents a retreat into contexts tractable with existing tools. To the contrary, we believe the original ambition of adaptive management to engender open-ended processes of social experimentation in complex situations, with scientists playing an active role in collaboration with managers and communities, is an important and valuable one. However, we suggest that realising the benefits of such open-ended engagements and embarking on more fruitful enactments of adaptive management rest upon recognising the multiple subjectivities underlying the ideas of nature embedded within adaptive management and the manifold natures produced through adaptive management practice. In the next section, we employ a performative lens to explore how subjectivities interact with technical practices in two Australian enactments of adaptive management.

Antipodean natures: the Australian fire context

Australia is the land of contrarieties, where the laws of nature seem reversed; her zoology can only be studied and unraveled on the spot, and that too only by a profound philosopher.

(Barron Field 1825, in Macinnis 2012, p. 4)

The temporal and spatial vagaries of Australian nature, including sporadic extremes of fire, drought and flooding, have long unsettled managers and scientists. Adaptive management has been enthusiastically embraced in Australian environmental management as a means to unravel these contrarieties. However, recognition that the 'laws of nature' referred to by Barron Field are, rather, the laws of a specifically European nature as known through a specifically European science points to the subjectivities that underlie the application of scientific rationality. To develop performative accounts of adaptive management, we need to unpack these subjectivities before we can understand the struggles of practitioners to enact it. As Griggs *et al.* note, institutions, policies and management paradigms are 'created and sustained by actors who, operating within large societal frames, traditions and hegemonic discourses, struggle with dilemmas and dislocations' (2014, p.11). We consequently situate our narratives of Australian adaptive management practice in historical context, because colonial social-ecological imaginaries continue to shape environmental management.

British colonials arrived in Australia in 1788 and were shocked by the 'new' landscapes and people (Anderson 2007). The contrarieties they experienced, from black swans to kangaroos, to Indigenous people who cropped in unfamiliar ways, simply did not fit the expected models (Carter 1987). Even as Barron Field noted that 'the laws of nature seem reversed', he and others were nonetheless certain that the four golden rules of the Enlightenment paradigm – order, reductionism, predictability and determinism, and knowledge equals order – would provide clarity (Geyer and Rihani 2010, pp. 13–14). It was not countenanced that this apparently timeless nature might be (a) subject to dynamic change and (b) the product of complex relationships between human practices and ecological processes over many thousands of years. Indeed, in a peculiar way, 1788 is the moment that Antipodean ecological time froze. The nature of 1788 frequently constitutes the 'desired state' that Australian adaptive management practitioners seek to 'adaptively' (re)produce today.

Nevertheless, the application of scientific rationality to the vagaries of the new-old continent was not a smooth process. The unfamiliarity of indigenous cropping practices and indigenous resistance to colonial civilisation meant that colonials were unable to conceive of Indigenous people as nature-altering beings (Anderson 2007, p. 199). This challenged the Enlightenment sense of self as outside nature and undermined explanations of nature made in terms of ordered hierarchies and universal laws. Indigenous non-conformity led to a revision of Enlightenment ordering to emphasise difference, separate speciation and the ranking of species (Anderson 2007, pp. 72–73). These dislocations highlight some crucial aspects of experience not captured in scientific rationality: the 'meaningful totality' of human–environment relationships in particular landscapes and the importance of context and temporality.

This is amplified in the Australian experience of fire. Fire occurs in an intrinsic relationship with the landscape. For instance, the most widespread fires in the Central Tanami Desert occur when above-average rainfall is followed by a dry period among particular vegetation communities such as spinifex grasses (Latz 2007, p. 82). Different natures have emerged from the interaction of these ecological processes and human activities over time, indicating the complex temporalities at play in attempts to manage fire adaptively. Indigenous Australians view fire as integral to the creation of nature and have used 'fire stick farming' over many thousands of years to provide sustenance and stimulate regeneration of vegetation (e.g. Jampijinpa *et al.* 2008, p. 31). However, Australian colonial foresters, trained in Europe, sought to suppress fire as a perceived threat to nature (Kessell 1920 in Burrows and McCaw 2013, p. 327). Today, fire stick farming is used to legitimate scientific modelling of contemporary ecological burning. The concept of 'managed burning', which lies at the core of an ecological burn, joins colonial ideas of Australian wilderness together with evocation of fire stick practices. In the following narratives, we consider the 'dilemmas and dislocations' faced by practitioners tasked with enacting adaptive fire management in two different landscapes and indicate where these practices might open up for more productive engagements with people and place.

Narrative #1: planned burning in Victoria

The environmental awakening in Australia from the 1950s onwards heralded new conservation values and helped revive an interest in fire as a means to create replicas of pre-1788 nature (Altangeril and Kull 2013). The prescribed burn is generally intended to protect humans and human assets and may involve, for instance, the creation of a firebreak by clearing trees and burning the vegetation in the designated area.

By contrast, an ecological burn is intended to imitate 'natural' outbreaks of fire. In ecological burns, mosaics of different heat intensity are produced in pursuit of particular ecological outcomes such as preserving tree hollows or managing invasive phalaris grass. Just as ecology has become a 'public science ...

Figure 12.1 Victoria planned burn.

Credit: K. Tolhurst.

central to government, bureaucracy and community' (Robin 1998, p. 3), adaptive management has become a preferred tool of ecology embraced by management organisations seeking to connect scientifically credible data from multiple sources to underlie an adaptive approach to fire management (Bradstock *et al.* 2012, p. 25).

However, the ecological representation of practice and purpose associated with nature-as-biodiversity in planned and ecological burning is frequently contested. A planned burn is part of a set of practices within a fire regime plan and is generally accepted if a planned burn in an ecosystem is primarily about preventing a wildfire and secondarily about maintaining ecosystem processes (Altangeril and Kull 2013, pp. 113–114). Adaptive management in this scenario is about monitoring and assessing the levels of vegetation as fuel, the climate cycles that affect the moisture levels in that potential fuel, and likely ignition sources if the fire is not deliberately set by managers. Knowledge of fire behaviour is based on small-scale trial-and-error experiments in the lab or field, extrapolated to larger spatial scales and to incorporate 'complex interactions between fire and the atmosphere' (Bradstock *et al.* 2012, pp. 16–17). Yet, with regard to biodiversity and functioning ecosystems, 'models [are] largely untested at spatial and temporal scales that are relevant to management' (Bradstock *et al.* 2012, p. 17). Therefore, the optimisation of biodiversity and the minimisation of wildfire spread frequently create competing goals. Details of desired landscape mosaics are rarely specified and there are few guidelines for practitioners (Penman *et al.* 2011, p. 727).

Consequently, uncertainty surrounds the likely success of a planned or ecological burn, and success may depend on defining the burn objectives as much as the physical management of the burn site. The ambiguity around the floral, faunal and soil data in most Australian ecosystems means many species are known only as they respond to fire, creating a particular version of nature through the accumulated fire regime (Gill *et al.* 2013, p. 441) rather than as an expression of a known or imagined nature, based on scientific evidence or social expectations.

If 'wild nature' spawns wild fire and lurks somewhere in the Australian landscape, it is most strongly perceived as an issue for public land[1] management of protected areas. On 7 February 2009, Victoria experienced a confluence of bushfires across public lands that killed 173 individuals as it spread to private properties, eventually affecting about 430,000 hectares (Attiwill and Adams 2013, p. 48). An investigation of the policy and practice response in the Victorian Bushfires Royal Commission (Parliament of Victoria 2010) led to a series of recommendations that included a 5–8 per cent annual burns target for public land across the state (Recommendation 56). Recommendation 56 was founded on the idea that planned burns reduce fuel and therefore the intensity of subsequent fires (Penman *et al.* 2011). The 5–8 per cent target originated in modelling of the eucalypt forests in the southwest of Western Australia – on the opposite side of the continent to Victoria – over 46 years (Attiwill and Adams 2013, p. 51). The application of the 5 per cent target is contested among fire scientists, with some considering it a minimum requirement for planned burns, others as an 'interference with nature' (Attiwill and Adams 2013, p. 51). This social-ecological imaginary suggests that 'ecological imbalances' could be overcome at a landscape scale by mimicking 'more natural fire regimes' (Jurskis 2005, p. 260). Bradstock *et al.* note that one consequence is a 'default position' in management that 'pyrodiversity favours biodiversity' (2012, p. 317).

The socio-political intent of the Victorian Bushfire Royal Commission was to protect human lives and assets by controlling fire (using planned burns to reduce fuel), particularly in peri-urban areas, but in performance the 5 per cent target was mostly directed at more remote public lands. Further, despite concessions to adaptive management in the presentation of the policy, the expected scientific enactment reflected a command-and-control approach. The separation of fire treatments as distinct variables in the landscape served to separate practice from science. In fire science the landscape easily becomes reduced to a hazard to be controlled and understood as wind speed, humidity and fuel load; whereas land managers' experience of these landscapes in non-fire season is, for example, forest management or tourism (Attiwill and Adams 2013, p. 49). The practitioners implementing the Victorian planned burns experienced at least three dislocations – first, that the modelling was historic, based on a place and climate different from Victoria; second, that planned burns are contested within science because of different social-ecological imaginaries about the interaction of fire and landscape and the versions of nature that ensue (Bradstock *et al.* 2012, p. 311); and third, that political trepidation about sparking uncontrollable fires actually prevented planned burns from taking place in those areas closest to human settlement (thus undermining the central justification for the policy). The performance of planned burns isolated fire treatments as discrete entities that framed and solved a distinct problem in response to a particular version of nature.

The Victorian Bushfire Royal Commission illustrates the social, technical and political imperatives that enmesh planned burns and frame the 'choreographies of practice' through which adaptive management is enacted (Griggs *et al.* 2014). A performative lens reveals that adaptive management in this case is less about the progressive reduction of uncertainties through decision choices and more about in-the-moment negotiations between dynamically intersecting temporalities. Despite awareness of the contradictions inherent in the scientific practices associated with fire management, the scientific argument is that improved fire-adaptive management depends on 'the degree to which we can control or domesticate fire' and that 'new ways of viewing, measuring and mapping fire' are needed (Bradstock *et al.* 2012, p. 319). In the next section, we examine how an indigenous mapping approach may offer guidance for the future development of adaptive management practice.

Narrative #2: the women's mapping project in the Tanami Desert

Parts of Warlpiri Country, within the Tanami Desert in the Northern Territory, Australia, are currently designated as IPA, otherwise known as IPA #23 (the Northern Tanami). IPA #23 was declared in 2007 and encompasses 40,000 square kilometres ranging from arid to sub-tropical landscapes (http://www.clc. org.au/articles/info/indigenous-protected-areas). IPAs are intended to promote sustainable use of resources, including conservation of biodiversity and cultural values. This emphasis on human presence differs from conventional protected area management which, for example, excludes residential occupation. The designation of IPA #23 has initiated a range of environmental management projects in partnership with local indigenous communities. In this narrative, we follow the women's mapping initiative – as described by J.P., a land protection officer with the Central Land Council – as part of a cross-cultural fire management program in the Central Tanami.

Warlpiri women are represented on the IPA committee and more lately in ranger groups (J.P., pers. comm. 2015). They are often less active participants in meetings, however, and the mapping initiative fulfils the interests of Walpiri women to go on country and convey their cultural knowledge to younger generations, with fire practices implicit in this interest. Gabrys and Vaarzon-Morel (2009) reviewed indigenous burning practices in the Central Tanami and noted the traditional differentiation between burning done by men and women. For example, women are said to burn around waterholes and in pursuit of small animals, and senior women must be involved in the management of women's sacred sites. The initiative sits within a broader adaptive management land use programme that combines 'traditional and contemporary fire knowledge, practices

Figure 12.2 Only a glimpse.
Credit: M. Erglis.

and technologies in annual cycles of planning, implementation, monitoring and review' (Broun and Allan 2011, p. 67). But the women's initiative avoids scientific rationality alone, focussing on the experience of practice, the relationships and structures inherent in action and as part of place. The women's approach to fire management refutes the subject–object duality in science, instead invoking a sense of 'being in the world' (compare Heidegger 1996) where nature is part of what it means to be human.

The adaptive management practices in the mapping initiative emerge from what the women say they need to do and are part of a process to facilitate cross-cultural management. As Nungarrayi, a senior Warlpiri woman from Yuendumu explains in relation to the planning process for the Southern Tanami IPA, 'we don't want *kardiya* [non-indigenous people] to come in with their own picture already painted about how it will happen … we need to sit down … and paint that picture together' (Preuss and Dixon 2012, p. 3). Management emerges as a practice grounded in doing and is not separate from the cosmology that informs the women's being in the world. To paraphrase Warlpiri, 'the people look after Country and the Country looks after us'. Country embodies the mutual reciprocity of skin, language, law and ceremony in the Warlpiri cosmology (Jampijinpa *et al.* 2008). On the journey described by J.P., the women ask the IPA facilitators to 'leave space' in the work agenda, and this opens up time for a sit-down in a place, from which emerges all manner of spontaneous practices that are determined by what is found in the place at the time of the visit. While project participants agree that there are important places to be visited and probably burned, the details of these decisions are only made while on country, where all participants experience the conditions first-hand. Things happen: hunting, burning, walking through country, stories on country. J.P. (pers. comm.) writes:

> [it] is always more than the whitefella notion of hunting in the first place. Warlpiri use the word 'wirlinyi' which means going away from one's camp and returning the same day. It's a process of traversal that sings up the relation with country and that responds actively to what is found there. There are always multiple practices occurring. It's not simply getting a goanna out of a hole, although this is part of it.

The Tanami landscapes of today have been shaped by the nineteenth- and twentieth-century dispossession of Indigenous people. Walker notes that 'traditional land management activities such as ceremony, burning, and hunting and gathering maintained the diversity and health of these landscapes, and their cultural significance (2010, p. 65). The Tanami can therefore be considered an anthropogenic landscape reliant on Aboriginal management. Jampijinpa *et al.* (2008, p. 25) explain: 'Warlpiri Law is the management of country' and 'it includes what westerners often call "traditional ecological knowledge"'[2] (p. 26). However, when indigenous communities were allowed to return to their country, they were discouraged from burning because of threats to neighbouring pastoral properties and infrastructure (Latz 2007). As the Indigenous Protected Area and Central Land Council reestablish some forms of burning based on indigenous technical knowledge in 'two-way' programs, they 'integrate indigenous ecological and cultural knowledge in the conservation of Australia's natural and cultural assets' (SEWPac 2011 in Preuss and Dixon 2012. p. 3). But Holmes and Jampijinpa are clear that:

> At this scale, the distinction between IEK as a system of knowledge and other aspects of Warlpiri knowledge disappears. IEK is not a commodity in which facts are learned by rote and added to the toolbox of management techniques. Rather, IEK is a process that Warlpiri engage with throughout their lives. In this process, the systems of knowledge and behavior that sustain country also sustain people's lives.
>
>   (2013, p. 9)

This understanding affirms the importance of practice that emerges from everyday engagement with the landscape and *precedes* knowledge. It is not just information about where a particular species is but about being active in the creation of new practices and dynamic responses. Indeed, the importance of creating time and space in which new forms of practice may be generated on country have been identified as crucial for productive forms of collaborative adaptive management in the Tanami and beyond (Walker 2010; Preuss and Dixon 2012). This brings into focus Latz's observation that the sedimentary layers in parts of this desert indicate the historical existence of woodlands rather than the contemporary spinifex grasslands (2007, pp. 86–87). If this is correct, then Traditional Owners alive today would have no experience of other than spinifex landscapes or spinifex fire regimes. Their traditional fire knowledge does not represent care of an ecology associated with a precolonial landscape, but rather embodies a timeless connection to manage as acts within the present – as Jampijinpa *et al.* (2008, p. 2) state, to find 'the "timeless" principles of culture' and apply them now. This makes their practices 'new-old-different' and 'in place' (part of country) all at the same time.

Mapping has been used in other Tanami projects to identify ranger-preferred sites for ecological burns and to incorporate Traditional Owners' understanding of those locations (Gabrys and Vaarzon-Morel 2009). But here we consider how the map is co-produced, creating a mimic but different (compare Bhaba 1994) adaptive management decision tool. As such, the map retains Western influences and may appear 'authoritative and taken for granted' (Krupar 2015, p. 91) but for its reconstitution as a visible representation of multiple practices (embodying scientific and practice rationalities), wherein the social situatedness of practice redefines the map meaning. Conventionally the Traditional Owners locate traditional sites of vegetation, small animals and fire. The map begins as an object depicting historic fire actions and becomes the subject of people–place management interaction conveying the endlessness-of-time-as-now. The map acts to dislocate the power of scientific rationality by reorganising rather than ordering (singing up country), renaming (stories of a goanna hunt) rather than reducing to a numeric code, and replaces the map information (data) by an assembling of these practices, adding meaning to information and creating knowledge based on doing. In this way the map symbolises the messiness of an ideal adaptive management – one that learns by doing and acknowledges practitioners and place and all of these in the context of the 'eternally unfolding present'.

Conclusions

In this chapter, we have argued that the gap between adaptive management theory and practice has emerged in part because using management experience as a basis for knowledge and action is problematic when conceived within epistemologies of science. We suggest that the roots of adaptive management in ecological pragmatism provide a route to reimagine 'learning-by-doing' through epistemologies of practice and performance. By drawing out the ways in which the practices of adaptive management are shaped by moment-to-moment interaction between social-ecological imaginaries, technical procedures and political imperatives, among others, epistemologies of practice bring to life the messiness – but also the possibilities – of adaptive management as a means of navigating dynamically changing relationships between people and landscapes. We have presented two exploratory narratives that demonstrate some potential rewards of this shift in registers. 'Planned burning in Victoria' reveals the ways that epistemologies of science disguise the range of considerations involved in adaptive management decision making. Applying a performative lens to this process suggests that adaptive management is less about the progressive reduction of uncertainty through time and more about the generation of practice that is able to navigate dynamically intersecting temporalities. The 'women's mapping initiative' demonstrates how indigenous epistemologies may unsettle epistemologies of science and produce enactments of adaptive management that embrace the temporally contingent production of knowledge, context and practice. Together, both narratives disrupt the 'subject–object' gaze of scientific rationality by revealing the fluidity and co-constitution of subject–object and subject–subject relations in practice. These brief narratives indicate the potential value of in-depth performative case studies of adaptive management. Such accounts may prompt a shift from considering adaptive management to be primarily about adaptively seeking a 'true state of nature' towards an emphasis on developing new forms of knowing and doing in place. This would have significant implications for the role of science in adaptive management and environmental management, from the provision of 'external' advice based on the subject–object gaze to reflexive participation in the everyday performance of natures.

Acknowledgements

Many thanks to Julia Perdevich, who recounted the women's mapping narrative, the Landscape and Environmental Sociology team at the University of Melbourne, and the editors of this volume. Your comments, suggestions and discussions have improved this chapter immeasurably. Simon West's time on this project was funded by the Swedish Research Council and Mistra.

Notes

1 Approximately one-third of the state of Victoria is public land and includes 3.4 million hectares of state forest and 4.2 million hectares of conservation areas (ENRC 2008 in Altangeril and Kull 2013).
2 Also referred to as indigenous ecological knowledge (IEK).

References

Allan, C. and Curtis, A. 2005. Nipped in the bud: Why regional scale adaptive management is not blooming. *Environmental Management*, 36(3), 414–425.
Allen, C.R. and Garmestani, A.S. 2015. Adaptive management, in *Adaptive Management of Social-Ecological Systems*, edited by C.R. Allen and A.S. Garmestani. Dordrecht, the Netherlands: Springer, 1–10.
Allen, C.R. and Gunderson, L.H. 2011. Pathology and failure in the design and implementation of adaptive management. *Journal of Environmental Management*, 92(5), 1379–1384.
Altangeril, K. and Kull, C. 2013. The prescribed burning debate in Australia: Conflicts and compatibilities. *Journal of Environmental Planning and Management*, 56(1), 103–120.
Anderson, K. 2007. *Race and the Crisis of Humanism*. New York: Routledge.
Attiwill, P. and Adams, M. 2013. Mega-fires, inquiries and politics in the eucalypt forests of Victoria south-eastern Australia. *Forest Ecology and Management*, 294, 45–53.

Beilin, R. and Bohnet, I. 2015. Culture-production-place and nature: The landscapes of somewhere. *Sustainability Science*, 10(2), 195–205.

Berkes, F., Colding, J. and Folke, C. (eds). 2003. *Navigating Social-Ecological Systems: Building Resilience for Complexity and Change.* Cambridge, UK: Cambridge University Press.

Bhaba, H. 1994. *The Location of Culture.* London: Routledge.

Bradstock, R.A., Williams, R.J. and Gill, A.M. 2012. Future fire regimes of Australian ecosystems: New perspectives on enduring questions of management, in *Flammable Australia: Fire Regimes, Biodiversity and Ecosystems in a Changing World*, edited by R.A. Bradstock, A. Malcolm Gill and R.J. Williams. Collingwood, Australia: CSIRO Publishing, 307–324.

Broun, G. and Allan, G. 2011. Case study: Community-based fire management in the Tanami Desert Region of Central Australia, in *Community-Based Fire Management in Practice*, United Nations Food and Agriculture Organization, Annex 4, 67–78. Accessed 21 June 2015. Available at http://www.fao.org/docrep/015/i2495e/i2495e13.pdf.

Burrows, N. and McCaw, L. 2013. Prescribed burning in southwestern Australian forests. *Frontiers in Ecology and the Environment*, 11, e25–e34.

Carter, P. 1987. *The Road to Botany Bay.* London, UK: Faber and Faber.

Cook, N. and Wagenaar, H. 2012. Navigating the eternally unfolding present: Toward an epistemology of practice. *The American Review of Public Administration*, 42(1), 3–38.

Cote, M. and Nightingale, A. 2012. Resilience thinking meets social theory: Situating social change in socio-ecological systems (SES) research. *Progress in Human Geography*, 36(4), 475–489.

Curtin, C. and Parker, J.P. 2014. Foundations of resilience thinking. *Conservation Biology*, 28(4), 912–923.

Etienne, M., Du Toit, D.R. and Pollard, S. 2011. ARDI: A co-construction method for participatory modeling in natural resources management. *Ecology and Society*, 16(1), 44–51.

European Science Foundation. 2010. Landscape in a changing world: Bridging divides, integrating disciplines, serving society. *COST-ESF Science Policy Briefing*, 41, 1–16.

Feldbaum, A.A. 1961. Dual control theory, part I. *Automation and Remote Control*, 21(9), 874–880.

Folke, C. 2006. Resilience: The emergence of a perspective for social-ecological systems analyses. *Global Environmental Change*, 16(3), 253–267.

Gabrys, K. and Vaarzon-Morel, P. 2009. Aboriginal burning issues in the southern Tanami: tradition-based fire knowledge, in *Desert Fire: Fire and Regional Land Management in the Arid Landscapes of Australia.* Alice Springs, Australia: Desert Knowledge CRC, 79–186.

Geyer, R. and Rihani, S. 2010. From orderly to complexity science, in *Complexity and Public Policy*, edited by R. Geyer and P. Cairney. Oxford, UK: Routledge, 12–16.

Gill, A.M., Stephens, S.L. and Cary, G.J. 2013. The worldwide "wildfire" problem. *Ecological Applications*, 23(2), 438–454.

Gregory, R., Ohlson, D. and Arvai, J. 2006. Deconstructing adaptive management: Criteria for applications to environmental management. *Ecological Applications*, 16(6), 2411–2425.

Griggs, S., Norval, A.J. and Wagenaar, H. (eds). 2014. *Practices of Freedom: Decentred Governance, Conflict and Democratic Participation.* Cambridge, UK: Cambridge University Press.

Heidegger, M. 1996. *Being and Time.* Translated by J. Stambaugh. Albany, NY: State University of New York Press.

Holling, C.S. 1973. Resilience and stability of ecological systems. *Annual Review of Ecology and Systematics*, 4, 1–23.

Holling, C.S. 1978. *Adaptive Environmental Assessment and Management*. London, UK: John Wiley.

Holling, C.S. and Meffe, G.K. 1996. Command and control and the pathology of natural resource management. *Conservation Biology*, 10(2), 328–337.

Holling, C.S. and Sundstrom, S.M. 2015. Adaptive management, a personal history, in *Adaptive Management of Social-Ecological Systems*, edited by C.R. Allen and A.S. Garmestani. Dordrecht, the Netherlands: Springer, 11–25.

Holmes, M.C.C. and Jampijinpa, W.(S.P.). 2013. Law for country: The structure of Warlpiri ecological knowledge and its application to natural resource management and ecosystem stewardship. *Ecology and Society*, 18(3), 136–148.

Jacobson, C., Hughey, K.F.D., Allen, W.J., Rixecker, S. and Carter, R.W. 2009. Toward more reflexive use of adaptive management. *Society and Natural Resources: An International Journal*, 22(5), 484–495.

Jacobson, S.K., Morris, J.K., Sanders, J.S., Wiley, E.N., Brooks, M., Bennetts, R.E., *et al.* 2006. Understanding barriers to implementation of an adaptive land management program. *Conservation Biology*, 20(5), 1516–1527.

Jampijinpa, W.(S.P.), Holmes, M. and Box, (L.)A. 2008. *Ngurra-kurlu: A Way of Working with Warlpiri People*, DKCRC Report 41. Alice Springs, Australia: Desert Knowledge CRC.

Johnson, B.L. 1999. Introduction to the special feature: Adaptive management – scientifically sound, socially challenged? *Conservation Ecology*, 3(1), 10.

Johnson, F.A., Williams, B.K. and Nichols, J.D. 2013. Resilience thinking and a decision-analytic approach to conservation: Strange bedfellows or essential partners? *Ecology and Society*, 18(2), 27.

Jurskis, V. 2005. Decline of eucalypt forests as a consequence of unnatural fire regimes. *Australian Forestry*, 68(4), 257–262.

Krupar, S. 2015. Map power and map methodologies for social justice. *Georgetown Journal of International Affairs*, August 10, 91–101.

Latz, P. 2007. *The Flaming Desert: Arid Australia—A Fire Shaped Landscape*. Alice Springs, Australia: Peter Latz.

Macinnis, P. 2012. *Curious Minds: The Discoveries of Australian Naturalists*. Canberra, Australia: National Library of Australia.

Moore, A., Hauser, C. and McCarthy, M. 2008. How we value the future affects our desire to learn. *Ecological Applications*, 18(4), 1061–1069.

Olwig, K. 1996. Recovering the substantive nature of landscape. *Annals of the Association of American Geographers*, 86(4), 630–653.

Parliament of Victoria. 2010. Victorian Bushfire Royal Commission. Summary. Accessed 21 July 2015. Available at http://www.royalcommission.vic.gov.au/finaldocuments/summary/HR/VBRC_Summary_HR.pdf.

Penman, T., Christie, F., Andersen, A., Bradstock, R., Cary, G., Henderson, M., *et al.* 2011. Prescribed burning: How can it work to conserve the things we value? *International Journal of Wildland Fire*, 20(6), 721–733.

Preuss, K. and Dixon, M. 2012. 'Looking after country two-ways': Insights into Indigenous community-based conservation from the Southern Tanamai. *Ecological Management and Restoration*, 13(1), 2–15.

Robertson, M. 2006. The nature that capital can see: Science, state and market in the commodification of ecosystem services. *Environment and Planning D: Society and Space*, 24(3), 367–387.

Robin, L. 1998. *Defending the Little Desert: The Rise of Ecological Consciousness in Australia.* Melbourne, Australia: Melbourne University Press.

Sandberg, J. and Tsoukas, H. 2011. Grasping the logic of practice: Theorizing through practical rationality. *Academy of Management Review*, 36(2), 338–360.

Scoones, I. 1999. New ecology and the social sciences: What prospects for a fruitful engagement? *Annual Review of Anthropology*, 28, 479–507.

Stirling, A. 2014. *Emancipating Transformations: From Controlling 'the Transition' to Culturing Plural Radical Progress.* STEPS Centre Working Paper 64. Brighton, England: STEPS Centre.

Wagenaar, H. 2011. *Meaning in Action: Interpretation and Dialogue in Policy Analysis.* Armonk, NY: M.E. Sharpe.

Walker, B. and Salt, D. 2006. *Resilience Thinking: Sustaining Ecosystems and People in a Changing World.* Washington, DC: Island Press.

Walker, J. 2010. *Processes for Effective Management: Learning from Agencies and Warlpiri People Involved in Managing the Northern Tanami Indigenous Protected Area, North Australian Institute of Advanced Studies.* Alice Springs, Australia: Charles Darwin University.

Walters, C. 1986. *Adaptive Management of Renewable Resources.* Caldwell, NJ: Blackburn Press.

Walters, C.J. and Hilborn, R. 1978. Ecological optimization and adaptive management. *Annual Review of Ecology and Systematics*, 9, 157–188.

Walters, C.J. and Holling, C.S. 1990. Large-scale management experiments and learning by doing. *Ecology*, 71(6), 2060–2068.

West, S. 2015. *Negotiating Social-Ecological Fit through Knowledge Practice.* Licenciate Thesis in Natural Resources Management. Stockholm, Sweden: Stockholm University.

Wiens, J.A. and Milne, B.T. 1989. Scaling of 'landscapes' in landscape ecology, or, landscape ecology from the beetle's perspective. *Landscape Ecology*, 3(2), 87–96.

Williams, B.K. 2011. Passive and active adaptive management: Approaches and an example. *Journal of Environmental Management*, 92(5), 1371–1378.

Williams, B.K. and Johnson, F.A. 2015. Optimization and resilience in natural resources management, in *Adaptive Management of Social-Ecological Systems*, edited by C.R. Allen and A.S. Garmestani. Dordrecht, the Netherlands: Springer, 217–234.

13 How to bring historical forms into the future?

An exploration of Swedish semi-natural grasslands

Marie Stenseke, Regina Lindborg,
Simon Jakobsson and Mattias Sandberg

Introduction

Semi-natural grasslands are one of the most species-rich habitats in Europe, hosting a significant flora and fauna of invertebrates, birds and mammals (Pykälä 2000; Lindborg *et al.* 2008). These grasslands are characterised by unsown vegetation; they have not been substantially modified by intensive fertilisation, drainage or herbicide use; and are maintained by livestock grazing. They also promote ecological functions that are of great importance for agriculture, for example, pollination and biological pest control (Beilin *et al.* 2014; Jonsson *et al.* 2014). In addition, in Sweden, the beauty and floral splendour of these habitats is a central part of the cultural heritage, since the biological and cultural qualities are the outcome of human interaction with non-human beings and processes. For a long time, pastures were a major feature in the agricultural landscape, but due to a rapid decrease in size as well as number, following structural changes in Swedish agriculture, they have been recognised as preservation sites since the 1980s (Swedish EPA 2012). The requested management of the grasslands, as pronounced in various schemes, entails a number of issues closely associated with time, such as land use history, provenance and restoration. The need for continuous human interactions demands reflection on how physical and biological features and processes originating from functions and contexts that no longer exist or have been marginalised (compare Lowenthal 1985; Rotherham 2013) link to contemporary cultural and societal contexts and processes. Temporal aspects are, however, often unarticulated in policies and management strategies. In this chapter we explore the issue of temporality in semi-natural grassland management and reflect on the implications of a heightened recognition of time for nature conservation strategies. Since time is an inherently fundamental question that has received relatively little attention in previous research on landscape management and nature conservation, we take a rather eclectic approach in order to stimulate further discussions.

Semi-natural grasslands in Sweden

Farming has been an active force on the Scandinavian Peninsula for 6,000 years (Eriksson *et al.* 2002). The first farming activities adopted were nomadic animal husbandry. Later, slash-and-burn agriculture was introduced, and around 2000 BP, permanent settlements were established, which included keeping cattle in stables during winter, thus providing manure for permanent crop fields. The combination of animal husbandry and cropping that became the way of farming in Scandinavia was well established in the Middle Ages, when Sweden emerged as a nation. Over the centuries, until the introduction of commercial fertilisers around AD 1900, the landscapes surrounding agricultural settlements were dominated by meadows for hay, pastures and forests used for grazing, while arable fields were rather small (Jansson 2011). Hence, the semi-natural grasslands have historically been of great importance to Swedish agriculture as grazing lands and fodder resources (Stenseke and Berg 2007).

Throughout the twentieth century, the area of semi-natural grasslands has considerably diminished in Sweden. Today these lands cover about 450,000 hectares, which is only some 10 per cent of their original extent (Eriksson *et al.* 2002). The main reason for this decline is that these grasslands do not easily fit into modern agriculture, being in general less productive and less appropriate for rationalisation due to physical landscape constraints (Queiroz *et al.* 2014). In the early transition phase, some of the former grasslands were transformed into arable land, but a large area has been abandoned and become overgrown or has been forested on purpose.

During the late twentieth and early twenty-first century, a number of national environmental aims concerning the preservation of semi-natural grasslands have been formulated, and policy measures have been introduced. In Sweden, similar to other European countries, the integrated relationship between nature and culture in farmland has been acknowledged in nature conservation management and planning, and animal husbandry is commonly accepted as a practice (Emanuelsson 2009). Management of almost 100 per cent of the identified semi-natural grasslands in Sweden is supported by subsidies from environmental schemes within the Common Agriculture Policy (CAP) (Swedish Government 2001, Swedish Board of Agriculture 2000). The majority of them are managed within the framework of private farms, while just a few are located in national parks or nature reserves. Grazing domestic animals is, thus, regarded as a means of nature conservation, with cows or sheep as key species in biodiversity.

The inclusion of semi-natural grasslands in the nature conservation agenda has challenged the overall framing of nature conservation. Here, we would like to point to two significant difficulties; first, nature conservation management has to be run in tandem with cultural heritage management, since the grasslands manifest both rich biodiversity and the history of human activities. In Swedish landscape management, the term bio-cultural heritage (in Swedish: biologiskt kulturarv) is frequently used and defined as 'ecosystems, land types and species that have emerged, evolved or been enhanced by human activities and whose long-term survival and development depends on or is positively influenced by land use and management' (Swedish National Heritage Board, 2014, p. 3). Notably, this definition differs slightly from the widespread international definition of bio-cultural heritage, in which the term is limited to lands linked to indigenous people (IIED 2015). For example, the integration of agricultural land use and conservation in Sweden contrasts with the separationist approach in Australia, where cows and sheep are not considered to belong, since nature is usually thought of as plants and animals that existed before European colonisation (House *et al.* 2008; Saltzman *et al.* 2011). Second, the biological qualities are dependent on continuous management, in contrast to a common nature conservation strategy of development without human interference. Hence, the recognised need for human intervention to maintain the qualities of semi-natural grasslands marks an approach divergent from the wilderness-biased nature conservation profile, in which 'free development', that is, no managerial interventions, is a general meas-ure and still dominant in other types of areas, for example, marine areas, alpine areas, forests and wetlands (Mels 1999). The distinctness of semi-natural grass-lands could be related both to the historical roots of Swedish nature conservation originating from romantic ideas about nature at the end of the nineteenth century and to the fact that biodiversity richness in agricultural landscapes has only been taken into account relatively recently. The very term 'semi-natural grasslands' (in Swedish: naturbetesmarker, which literally means natural pasturelands) sig-nals that the characteristics of these lands differ from 'pure nature' compared to other nature conservation target areas. To illustrate: Ängsö National Park was an early example of the implications of the contrasting managerial approaches. Ängsö is an island in the Stockholm archipelago, and the national park was inau-gurated in 1909 because of the rich flora on grazed pastures and mowed hay meadows. In line with the wilderness paradigm in nature protection at that time, agricultural activities were forced to cease. After a couple of decades, the rich biodiversity was severely threatened by overgrowth. Hence, grazing animals and mowing were reintroduced (Wästfelt *et al.* 2012).

Both these challenges, the former relating to the past and the latter to the future, demonstrate the dynamic character of landscapes and the necessity of considering the temporal dimension when considering the effectiveness of the measures taken to maintain the identified qualities in semi-natural grasslands. We want to illustrate this by drawing on an interdisciplinary study in the Östra Vätterbranterna (in English: East Vättern Escarpments) in south-central Sweden. While the overarching aim of the study is to contribute to the development of sustainable land use strategies, our focus in this chapter is on nature conserva-tion aims and measures. Our concern is not primarily to describe the dynamic character of landscapes as such. Hence, we distinguish *time* in the sense of the sequence of events related to ongoing processes, such as ageing, seasonality and dispersal of ideas, from *temporality*, here understood as how time is perceived and addressed in strategies and practices concerning semi-natural grasslands (compare Ingold 1993). In the following, we present a processual landscape perspective, serving as a theoretical framework. We introduce the study area, Östra Vätterbranterna, and elaborate on a number of time-related issues con-cerning landscape management strategies and practices in the area. The chapter ends with some reflections on how temporal aspects could be better addressed in policies and management strategies.

A processual landscape perspective on semi-natural grasslands

Elaboration of the temporal dimensions in semi-natural grassland management requires a framing that can integrate both time and space to understand the dynamics of landscapes. The material dimension of landscapes, that is, the morphology and the physical content of the grasslands, including plants, soils, stones and artefacts, is in focus in this chapter; however, a broader understanding of what the landscape 'is' and what the prerequisites are for its future state has to be recognised (O'Rourke 2005). In the social sciences and humanities focussing on human–nature relations, a processual and relational landscape perspective is widely acknowledged. 'Landscape' is then a concept to grasp the momentary 'thereness' and relative location of all entities (Sui 2012). It is conceived as a spatio-temporal continuum, consisting of coevolving material and immaterial features: physical elements, socio-economic features, institutional components and intangible aspects (Hägerstrand 2001; Wylie 2007; ESF 2010). This fits well with the dynamic multi-equilibrium perspective in ecology (Holling 1973), which has outdated the more static equilibrium view (Clements 1936). In terms of semi-natural grasslands, we must acknowledge that the composition of species is constantly changing and coevolving with non-human features and processes as well as features and processes related to human society.

In order to assist a structured understanding of landscape that connects time and space and helps to elucidate how landscapes evolve, Widgren (2004) has suggested a framework of four interrelated aspects: form, function, context and process. *Form* is the physical form, morphology and content. Generally, the aims of landscape management from both nature conservation and cultural heritage perspectives mainly concern this form. *Context* is the contemporary social and cultural context within which the physical features and structures in a piece of landscape exist, for example, the conditions for agriculture in Sweden and the socio-economic characteristics of Swedish society. *Function* relates to the meaning/s, the raison d'être that the specific forms have in their spatio-temporal context. Since it has been argued that only functional forms will endure (Antrop 2006; Rotherham 2013), the present and future functions of semi-natural grasslands are of interest.

The term *process* explicitly brings in the time dimension. It includes all kinds of ongoing changes, for example, biotic processes of growth, structural changes in the food market and societal changes of habits and values, and helps to recognise that the landscape at every given moment is a continuing result of a 'fabric of existence' (Hägerstrand 2009, p. 130; Beilin and West, this volume). In ecological research it has been shown that revealing ongoing processes is a critical issue. To date, research on landscape management and biodiversity has mainly focused on the effects of current spatial landscape structure on single species and species richness (Tscharntke *et al.* 2012). A number of crucial changes driving species loss have been identified, such as habitat deterioration, reduction of habitat area and increasing isolation of remaining habitats. Since there can be a time lag in the response, an extinction debt (Kuussaari *et al.* 2009), the conditions for species persistence might no longer be met by current landscape configuration or management, although the species are still present. Therefore, the number of endangered species may be underestimated. However, changing conditions may also give threatened species a chance to recover if a habitat is restored. Thus, having a spatio-temporal approach to conservation management may overcome the misleading interpretation that current landscape structure relates only to present-day landscapes. Since extinction debt, with time-delayed responses up to 50–100 years after land use change, has been documented in semi-natural grasslands for both plants (Lindborg and Eriksson 2004; Helm *et al.* 2006; Cousins 2009) and insects (Öckinger *et al.* 2012), future species loss may be expected even if the present landscape is maintained.

In sum, landscapes are in constant change, and successful landscape management implies not only the task of continuously revealing processes in the present and reassessing and elaborating on functions and contexts (compare Stenseke 2016) but also reconsidering what physical forms and content to aim at. Next we present the Östra Vätterbranterna, which is the empirical basis for exploring the time aspect in semi-natural grassland management. We have undertaken biological field inventories, interviewed farmers and executives, examined nature conservation documents, inventories, strategies, and documents related to the study area. In addition, we have also investigated related national strategies and policy texts.

Östra Vätterbranterna

Östra Vätterbranterna is situated along the shore of Lake Vättern in southern Sweden. The rift valleys that cut through this area have stimulated the development of distinct sub-regions regarding land use. Arable fields dominate valleys with deeper soils, whereas the hills have been mainly used for forestry and grazing. Intensified forestry practices from the mid-nineteenth century clearly affected the landscape, but it has retained a large amount of tree-rich semi-natural grasslands, that is, wood pastures that are still managed and form important landscape elements, compared to the rest of Sweden. While agriculture is based on dairy and beef production, the fact that it is compulsory to let the animals out during the grazing season in Sweden and that environmental subsidies are often of great economic importance combine to explain the continued management of the pastures. The size of the farms in the area varies from just a few hectares to 400 hectares, including forests, fields and semi-natural grasslands. Many of those managing the grasslands are part-time farmers or leisure farmers. The proximity to the city of Jönköping (90,000 inhabitants) makes it fairly easy to combine farming with other employment. It also means that the mosaic scenery of Östra Vätterbranterna is attractive for non-farming residents as well as for second homes, and the area serves as an important arena for outdoor recreation for the growing urban population.

Many of the current wood pastures in Östra Vätterbranterna consist of mosaic lands where small woody patches are surrounded by open grassland strips, which were formerly used for rotation agriculture including both crop production and grazing (Jakobsson and Lindborg 2014). The tree cover and the composition of

Figure 13.1 A wooded pasture in Östra Vätterbranterna.

Credit: Mattias Sandberg.

tree and shrub species vary a lot. There are often different species in a single pasture, but birch (*Betula pendula*), oak (*Quercus robur*), hazel (*Corylus avellana*) and rowan (*Sorbus aucuparia*) tend to dominate, and occasionally also pine (*Pinus sylvestris*). See Figure 13.1.

Old trees are important features of pastures, as well as habitats for lichens and insects. Moreover, wood pastures offer an environment essential for many nationally threatened species and species-rich environments in general. Each pasture has its unique properties, thus the between-site variation is a biodiversity trigger itself, for example, by attracting rare birds to breed, such as the Eurasian golden oriole (*Oriolus oriolus*), ortolan bunting (*Emberiza hortulana*) and common rosefinch (*Carpodacus erythrinus*). These wood pastures also sustain a very species-rich flora with species originating from forests, meadows and open pastures, many of which are otherwise rare in Sweden, such as field gentian (*Gentianella campestris*) and eyewort (*Euphrasia rostkoviana ssp. Fennica*).

Threats identified to the wooded semi-natural grasslands and their ecological status, mainly abandonment and becoming overgrown, stimulated a cooperative project including a range of stakeholders, and were one of the main reasons for the foundation of the Biosphere Reserve East Vättern Scarp Landscape in 2012. The establishment of the Biosphere Reserve has led to significant restoration and preservation efforts throughout the whole area (Jonegård *et al.* 2011). The label 'biosphere reserve' is a designation given by UNESCO's Man and the Biosphere (MAB) programme to areas that serve as good examples for sustainable development. There is a particular focus on the implementation of the Convention on

Biodiversity, while secondary aims are to promote social and economic development in tandem with the preservation of nature qualities and making the designated areas into learning sites (UNESCO 2015). Like UNESCO's world heritage sites, they are not protected per se, but they hold core areas of strictly protected ecosystems.

Landscape management and time

The landscape management strategies and practices in Östra Vätterbranterna evoke a number of time-related concerns. This section discusses three issues that have come to the fore in our project on landscape management in the area. We use the terminology of Widgren (2004), that is, form, context, function and process, as the basis of our discussion. First, we examine how the issue of time is addressed or framed in contemporary landscape management strategy documents. Restoration and re-creation, as nature conservation measures, are given particular attention. As a contrast to the strategic approaches, we then present and discuss findings from interviews with farmers and the time dimension in their practices. Finally, the issue of old trees in the pastures is used to draw a more detailed picture of the issue of temporality.

The objects and time frames of nature conservation strategies

Regional nature conservation strategy documents that concern semi-natural grasslands in Östra Vätterbranterna focus on species and structure, that is, the forms and physical content of the grasslands (Jonsson 2004; County Administrative Board of Jönköpings län 2006). A somewhat static view on landscape is indicated in the nature conservation strategy concerning Östra Vätterbranterna, conveyed in expressions such as 'typical nature qualities are prioritised', 'the area of the natural forest' and 'the original natural surroundings' (Jonsson, 2004, pp. 32, 48). Consequently, mapping species and monitoring how they thrive have been crucial, in addition to mapping habitats. The strategies for maintenance evolve around strengthening nature conservation itself as a function and include suggestions about area protection, subsidies and information. The long-term plans for, and effects of, subsidies and information are not elaborated in any detail or related to a larger societal context. Instead, they appear to be short-term solutions over the span of a few years. Overall, little is said about the relation to other functions, and the issues of context and processes that drive landscape change are largely ignored (County Administrative Board of Jönköpings län 2006).

The strategy of limiting the perspective to preserving the present state is consistent with the red list for Sweden, the national adaptation of the IUCN directives for red lists (Artdatabanken 2015; IUCN 2001), which serves as an important point of departure for the nature conservation approach in the area. Out of the five criteria that are the basis for the red list's categorisation of the extinction risk, there is a time dimension in two. Both of these refer to the present species composition:

(i) an observed, estimated, inferred or suspected population size reduction of ≥15 per cent over the last 10 years or three generations, whichever is longer; and

(ii) quantitative analysis showing the probability of extinction in the wild is at least 5 per cent within 100 years (Artdatabanken 2015).

Similarly, on a more general level, the Swedish environmental quality objectives, which serve as a basis for environmental policies and strategies, also signal a less dynamic view of the physical environment, especially when it comes to future prospects. There is an overall generation goal, implying that the objectives describe the desired state by 2020 (Swedish EPA 2012), based on a somewhat implicit idea that the environmental problems are to be solved by then. This viewpoint is dubious, since for most of the environmental problems there can be no fixed solutions but rather continuous management concerns. If applied to the landscape-related objectives, including maintenance of semi-natural grasslands, the idea that environmental problems can be solved by a fixed date means disregarding the changing character of the physical environment; disregarding not only changes due to climate change and other human induced effects but also changes due to non-human processes. The objective of 'a varied agricultural landscape' includes specifications related to semi-natural grasslands. Notably, these specifications have frequent references to the present state, and terms like 'preserve' and 'recover' are common (Swedish EPA 2012).

Generally, the environmental quality objectives as well as the red list relate to an equilibrium-centred ecology, despite the fact that this view is almost obsolete in science. Moreover, with the well-recognised prognosis for climate change, preservation of the present state is not likely to be a feasible alternative in a time perspective stretching beyond a couple of decades. One of the specifications to the environmental quality objective on agricultural landscapes states, 'Important ecosystem services of the agricultural landscape are preserved' (Swedish EPA 2013, n.p.), although this might be read as a less fixed view of the future landscape form and context. The environmental quality objective, explicitly related to biodiversity, 'a rich diversity of plant and animal life', also includes an opening for a more processual landscape perspective. It shows in the objective specification 'ecosystems have the ability to cope with disturbances and adapt to change, such as a changed climate, so that they can continue to provide ecosystem services and contribute to combating climate change and its effects' (Swedish EPA 2013, n.p.). In the case of Östra Vätterbranterna, however, this objective specification is not mirrored in the strategies.

Integrating nature conservation aspects of landscape management with cultural heritage concerns complicates the issue of what is to be maintained. In the national environmental quality objectives, one of the specifications concerning agricultural landscapes states, 'Biological and cultural heritage values of the agricultural landscape that have emerged through long-term, traditional management are preserved or improved' (Swedish EPA 2013, n.p.). 'Traditional' is a frequently used time-related term, but it is not obvious what traditional means in a landscape context (Eriksson 2011). From a nature conservation perspective, it is usually the character of the physical form, that is, structures and features with some history, whereas from a cultural heritage perspective, not only the forms but also the practices (e.g. mowing with scythes) and the functions (e.g. food production) might be included. This has implications for which activities are sanctioned and subsidised (compare Eriksson 2011). Furthermore it has to be recognised that intentional landscape management, that is, strategies and practices explicitly aimed at maintaining certain forms, has a history and, moreover, is a vital aspect of the contemporary relation between humans and especially the rural landscape (compare Cooke, this volume). In the area of Östra Vätterbranterna, efforts to protect and preserve certain landscape features started several decades ago and have been an important part of the process resulting in the present landscape forms. Through subsidies and payments, landscape management has become an emerging business (Wästfelt *et al.* 2012).

Restoration – linking the past to the future

The practices in landscape management of restoring and recreating are of particular interest for the issue of temporality. Two crucial time-related questions can be identified:

(i)   What is to be restored?
(ii) How can the effects of restoration be maintained, thus making the act of restoring successful in the long term?

Concerning the links to the past, the main goal in classic ecological restoration was to recreate wilderness, a state similar to what existed before human settlement (Hilderbrand *et al.* 2005). This was built on the Clementsian 'balance of nature' paradigm where plant succession terminates in a climax community, which remains at equilibrium until disturbed, after which the process of succession is restarted until the climax is reached. In order to succeed, ecological restoration must be framed at relevant spatio-temporal scales, which may be defined temporally with reference to ecological processes such as disturbance regimes and spatially with reference to ecological units such as demarcated areas and ecosystems (Suding *et al.* 2004; Bullock *et al.* 2011).

As for the future, it is problematic that the goals are unclear in many restoration projects, making it difficult to define whether or not a restoration is 'successful' (Drayton and Primack 2012). The outcome of biodiversity restoration is often measured as an increase in overall species richness or in rare species (Pykälä 2003; Piqueray *et al.* 2011); however, measurable goals of historical and aesthetic values are perceived as more problematic (Hansson and Fogelfors 2000; Lindborg and Eriksson 2004); they are more subjective and cannot be seen as independent from the spatio-temporal context (Lowenthal 2005). The success of restoration efforts should not just be measured against the state when the restoration activities are completed, but should also include the prospects for maintaining the results. The issue of maintenance is, however, absent from the strategies for the semi-natural grasslands in Östra Vätterbranterna. There seems to be an implicit reliance on continued public financial support for conservation measures, since the benefits of ecological restoration are discussed in terms of enhancing the functions of rich biodiversity and ecosystem services (compare Bullock *et al.* 2011; Benayas and Bullock 2012).

Farmers and the time dimension in landscape management

Farmers are the primary actors in current rural landscape management. In the foreseeable future, without farmers and grazing animals there will be very few semi-natural pastures left to protect. Therefore, the broader context in which these farmers act, giving the semi-natural grasslands and their maintenance a function in their lives and projects, is of vital interest (Stenseke 2006). Notably, the economic yield for many full-time and part-time farmers in Sweden relies on subsidies within the CAP. Hence, landscape management and restoration have an income-generating function for the farmers, and the directives for economic support within the policy are of vital importance for their practice.

When it comes to time, farming is seasonal in its nature and very much so at the northern latitudes where Sweden is geographically situated. The daily practice on the farm is, furthermore, dependent on the temporal rhythms of both human and non-human entities and processes (Peterson 2006). The different rhythms involved do not necessarily create harmony. While modern society is guided by time as an abstract and freestanding entity, shown by clocks and calendars (Sui 2012), farmers must still synchronise their actions with the needs of other living creatures, which by no means automatically match the clock or the day of the week (compare Lockie 2006). Thus, behind the picturesque and perhaps timeless appearance of the semi-natural grasslands lies devotion and many hours of labour.

The crucial importance of time, or rather timing, for the farmers in Östra Vätterbranterna in keeping animals and contributing to conservation is well illustrated by the situation for one of the largest milk producers in the area. This farmer recently expanded the farm business, made large investments in milking robots and doubled the numbers of cows. The expansion relies on access to both arable fields and semi-natural grasslands for grazing (and working off-farm one day a week). In order to have enough land to feed the cows, the farm structure is by necessity rather scattered in this topographically varied area, made up of small farmsteads. Looking after the animals on a daily basis, the transport of animals to different small pastures and the maintenance of fences consumes both time and money. Hence, the landscape form, once functional in co-creating the rich flora and fauna, has become more and more of an obstacle for future (intensive) farming prospects in the area. Still, the farmer expresses appreciation of the beauty of semi-natural grasslands and would like to spend even more time on conservation measures, for instance, restoring former pastures.

For small-scale landowners, the semi-natural grasslands often play a different role. Landscape maintenance is commonly regarded as a way to relax and exercise, a meaningful pastime. The off-farm income makes them less dependent on the profit from their land and implies more freedom to work with things they enjoy. Being under less economic pressure than their full-time farming neighbours allows leisure farmers and part-time farmers to engage in different conservation measures to a greater extent for their own interest and enjoyment.

Old trees in semi-natural grasslands

Trees are very much a focus for nature conservation interests in Östra Vätterbranterna. Two features of specific interest related to semi-natural grasslands are giant trees and pollarded trees (i.e. trees from which the upper branches are regularly removed), due to their importance as habitats for a diversity of lichens, birds, insects and fungi (see Figure 13.2). The pollarded trees also link back to practices in the farming system up to a century ago, when the leaves served as important fodder resources. From a nature conservation perspective, a lack of old trees is recognised, and pollarding is put forward as an important conservation measure, in terms of restoration pollarding, new pollarding and 'veteranising', that is, forcing the ageing process. By cutting off branches the tree is stressed, and this creates holes and wood is laid bare, which in turn attracts species that depend on old trees and dead wood.

In nature conservation documents for Östra Vätterbranterna, the history of the old trees is well recognised. Due to their age, these trees ensure a perspective of several centuries. They have a variety of ecological and cultural functions during the course of their lifetimes, and even dead wood is highly valued from a conservation point of view (Gränna skogsgrupp, n.d.). When it comes to maintenance, however,

Figure 13.2 A pollarded ash tree.

Credit: Simon Jonegård.

there is a focus on the present state, while long-term prospects are rarely addressed. Even though some consideration for the future provision of old trees and dead wood is expressed, it is rarely put into a wider context or discussion about new functions and uses to regenerate new trees with equivalent values. Furthermore, the strategy for valuable trees (County Administrative Board of Jönköpings län 2013) deals with actions on a micro-level, which include how to manage specific trees and their close environment. A number of threats towards the trees in Östra Vätterbranterna have been identified, with overgrowth following farmland abandonment seen as the most serious. Other risks are related to cuts due to land exploitation or farm rationalisation, discontinued pollarding and diseases. In the strategy documents for handling the threats, there is, however, little discussion of the societal context, changed functions, ongoing societal processes and how to address these. The only function that is extensively considered is biodiversity conservation. Even though information about the nature conservation values of the trees and wider cooperation between various stakeholders are mentioned as means, these ideas are not elaborated upon.

One may questioned whether intentionally (re)created landscape features are as valuable as 'organically' evolved ones. From an ecological point of view, trees that have recently been pollarded are generally valuable, since the rich biodiversity makes it valid to enhance a magnitude of species, wherever the trees are found in the area. In cultural heritage, history is a crucial dimension, implying that the age of a feature is important for how it is valued. There is also a deeply rooted recognition of provenance, implying an inherent quality in direct bonds to previous generations and their activities. Furthermore, cultural heritage is, with a few exceptions, place-specific and also related to a particular cultural context. New pollarding, thus, might not necessarily be motivated by cultural heritage objectives. Nevertheless, it is important to recognise that there might be other societal values of the newly pollarded trees, related to new functions such as their perceived beauty in the present context.

The farmers interviewed were in agreement about which trees they want to save on the semi-natural grasslands. The beauty of the trees is put forward as the most distinct argument. One of the youngest farmers interviewed has pollarded around 80 trees together with a neighbour. He is doing it 'for fun', along with the regular clearance of trees and bushes on the semi-natural grasslands. Since the leaves are not used as fodder, the form, that is, the aesthetic of the pollarded trees, gives them a function for the farmers. Moreover, the work itself, creating something considered beautiful, also gives meaning and a sense of accomplishment.

Reflections

The dynamic character of landscapes implies that time is a vital dimension in landscape management. Our exploration of temporal aspects and how they are addressed in public documents on semi-natural grassland management in Östra Vätterbranterna shows that the strategies are heavily concentrated on preserving the present state and what are perceived as acute needs in order to prevent further loss in the number of individuals of vulnerable species and ecosystems. While the landscape management strategies are clearly related to land use history, the connection to present functions and processes is less clearly acknowledged. Moreover, it appears to be a significant challenge to address the future, meaning some 20 years and more ahead. The lack of long-term forward-looking perspectives is already problematic from a strict ecological perspective, due to possible extinction debts. Insufficient consideration of the ever-changing socio-cultural context, of the issue of function on a societal as well as a farm level and of the ongoing processes that drive change, amplifies the problem. Hence, the strategies might be inefficient or, at worst, counterproductive over a perspective of several decades. Consequently, we argue that a recognition of the time dimension implies a clearer differentiation between short-term objectives and long-term considerations. The present narrower strategies, to preserve what is there in the semi-natural grasslands, might well be necessary in order to guide our immediate actions, but they need to be complemented by and related to plans addressing the future some decades ahead.

The long-term perspective demands a broader approach, a higher degree of integration between nature conservation and various other societal aspects and interests (compare Robin, this volume). Through the nature conservation strategies formulated thus far, biodiversity conservation has replaced food production as the dominant function in a broader societal perspective. The case of Östra Vätterbranterna shows that there is a heavy focus on strengthening the conservation function of wood pastures, through environmental payments. At the same time, other functions exist on the societal level as well as among individual farmers, related to identity, beauty and recreation. In order to improve the prospects for semi-natural grasslands at Östra Vätterbranterna, including the wood-pastures, we argue that the potential in these other functions should be explored and considered. This does, however, require inter-sectorial cooperation. Recognising the landscape as a process, where the composition of species coevolves with features related to human society, there is little point in nature conservation making its own scoping of the future and search for long-term strategies. Instead, this should be done in the frame of a broad search for sustainable land use and sustainable pathways (compare Slätmo, this volume).

The biosphere reserve appears to be a promising approach, by bringing in a wider perspective into landscape management and by launching ideas about new functions and adaptations to the contemporary societal context, such as local labelling of products from the area, small-scale wood processing, tourism and energy production (Jonegård *et al.* 2011). Notwithstanding that biodiversity preservation is an overarching objective for biosphere reserves, one mission, stated by UNESCO, is 'the development and integration of knowledge, including science, to advance our understanding of interactions between people and the rest of nature' (UNESCO 2015). Hence, it also poses a challenge to researchers to theoretically and methodologically advance ways of bridging specialised knowledge traditions.

References

Antrop, M. 2006. Sustainable landscapes: Contradiction, fiction or utopia? *Landscape and Urban Planning*, 75(3–4), 187–197.
Artdatabanken. 2015. *Rödlistans Kriterier och Kategorier*. Accessed 6 May 2015. Available at http://www.artdatabanken.se/naturvaard/roedlistning/kriterier-och-kategorier/.

Beilin, R., Lindborg, R., Stenseke, M., Pereira, H.M., Llausàs, A., Slätmo, E., *et al.* 2014. Analysing how drivers of agricultural land abandonment affect biodiversity and cultural landscapes using case studies from Scandinavia, Iberia and Oceania. *Land Use Policy*, 36(January), 60–72.

Benayas, J.M.R. and Bullock, J.M. 2012. Restoration of biodiversity and ecosystem services on agricultural land. *Ecosystems*, 15(6), 883–899.

Bullock, J.M., Aronson, J., Newton, A.C., Pywell, R.F. and Rey-Benayas, J.M. 2011. Restoration of ecosystem services and biodiversity: Conflicts and opportunities. *Trends in Ecology and Evolution*, 26(10), 541–549.

Clements, F.E. 1936. Nature and structure of the climax. *The Journal of Ecology*, 24(1), 252–284.

County Administrative Board of Jönköpings län. 2006. *Trakter med Särskilt Värdefulla Ängs- och Betesmarker i Jönköpings Län – Ett Underlag Till en Regional Bevarandestrategi.* Report 2006:49, Jönköping.

County Administrative Board of Jönköpings län. 2013. *Strategi för Skyddsvärda Träd i Jönköpings län.* Report 2013:07, Jönköping.

Cousins, S.A.O. 2009. Landscape history and soil properties affect grassland decline and plant species richness in rural landscapes. *Biological Conservation*, 142(11), 2752–2758.

Drayton, B. and Primack, R.B. 2012. Success Rates for Reintroductions of Eight Perennial Plant Species after 15 Years. *Restoration Ecology*, 20(3), 299–303.

Emanuelsson, U. 2009. *The Rural Landscapes of Europe: How Man has Shaped European Nature.* Stockholm: Formas.

Eriksson, C. 2011. What is traditional pastoral farming? The politics of heritage and 'real values' in Swedish summer farms (fäbodbruk). *Pastoralism: Research, Policy and Practice*, 1(25), 1–18.

Eriksson, O., Cousins, S.O.A. and Bruun, H.H. 2002. Land-use history and fragmentation of traditionally managed grasslands in Scandinavia. *Journal of Vegetation Science*, 13(5), 743–748.

[ESF] European Science Foundation. 2010. Landscape in a changing world: Bridging divides, integrating disciplines, serving society. *Science Policy Briefing* 41, 1–15.

Gränna skogsgrupp. n.d. *Betesmarkers naturvärdesträd i Östra Vätterbranterna – Effekter av Jordbruksverkets 50-trädsregel för betesmarker med EU-baserade stöd i ett urval nyckelbiotoper i odlingslandskapet.* Accessed 17 June 2015. Available at http://www.grannaskog.se/Hagmark50Trad.pdf.

Hägerstrand, T. 2001. A look at the political geography of environmental management, in *Sustainable Landscapes and Lifeways: Scale and Appropriateness*, edited by A. Buttimer. Cork: Cork University Press, 35–58.

Hägerstrand, T. 2009. *Tillvaroväven.* Stockholm: Formas.

Hansson, M. and Fogelfors, H. 2000. Management of a semi-natural grassland: Results from a 15-year-old experiment in southern Sweden. *Journal of Vegetation Science*, 11(1), 31–38.

Helm, A., Hanski, I. and Pärtel, M. 2006. Slow response of plant species richness to habitat loss and fragmentation. *Ecology Letters*, 9(1), 72–77.

Hilderbrand, R.H., Watts, A.C. and Randle, A.M. 2005. The myths of restoration ecology. *Ecology and Society*, 10(1), 19.

Holling, C.S. 1973. Resilience and stability of ecological systems. *Annual Review of Ecology and Systematics*, 4, 1–23.

House, A.P.N., MacLeod, N.D., Cullen, B., Whitebread, A.M., Brown, S.D. and Mclvor, J.G. 2008. Integrating production and natural resource management on mixed farms in eastern Australia: the cost of conservation in agricultural landscapes. *Agriculture, Ecosystems and Environment*, 127(3), 153–165.

[IIED], International Institute for Environment and Development. 2015. Accessed 1 June 2015. Available at http://biocultural.iied.org/about-biocultural-heritage.

Ingold, T. 1993. The temporality of the landscape. *World Archaeology*, 25(2), 152–174.

IUCN. 2001. *Red list categories and criteria*. Version 3.1 Second edition. Accessed 20 June 2015. Available at https://portals.iucn.org/library/efiles/documents/RL-2001-001-2nd.pdf.

Jakobsson, S. and Lindborg, R. 2014. Wood pasture profile: East Vättern Scarp Landscape, Sweden, in *European Wood-Pastures in Transition – A Social-Ecological Approach*, edited by T. Hartel and T. Plieninger. Abingdon: Routledge, 162–163.

Jansson, U. (ed.). 2011. *Agriculture and Forestry in Sweden since 1900: A Cartographic Description, National Atlas of Sweden*. Stockholm: Nordstedts.

Jonegård, S., Uhr, J., Lindell, M., Lund, M., Lund, A., Wallander, A., *et al.* 2011. *East Vättern Scarp Landscape Biosphere reserve nomination form*, Jönköping: Swedish Forestry Agency, County Administrative Board of Jönköping, Municipality of Jönköping, Federation of Swedish Farmers, Södra Skogsägarna, Gränna Skogsgrupp and World Wildlife Fund WWF.

Jonsson, P. 2004. *Östra Vätterbranternas Naturvärden – Arter, Miljöer och Dellandskap.* County Administrative Board of Jönköpings län, Report 2004:35, Jönköping.

Jonsson, M., Bommarco, R., Ekbom, B., Smith, H.G, Bengtsson, J., Caballero-Lopez, B., *et al.* 2014. Ecological production functions for biological control services in agricultural landscapes. *Methods in Ecology and Evolution*, 5(3), 243–252.

Kuussaari, M., Bommarco, R., Heikkinen, R., Helm, A., Krauss, J., Lindborg, E., *et al.* 2009. Extinction debt: A challenge for biodiversity conservation. *Trends in Ecology and Evolution*, 24(10), 564–571.

Lindborg, R., Bengtsson, J., Berg, Å., Cousins, S.A.O., Eriksson, O., Gustafsson, T., *et al.* 2008. A landscape perspective on conservation of semi-natural grasslands. *Agriculture, Ecosystems and Environment*, 125(1), 213–222.

Lindborg, R. and Eriksson, O. 2004. Historical landscape connectivity affects present plant species diversity. *Ecology*, 85, 1840–1845.

Lockie, S. 2006. Networks of agri-environmental action: Temporality, spatiality and identity within agricultural environments. *Sociologia Ruralis*, 46(1), 22–39.

Lowenthal, D. 1985. *The Past is a Foreign Country*. Cambridge, UK: Cambridge University Press.

Lowenthal, D. 2005. *The Past is a Foreign Country*. Cambridge, UK: Cambridge University Press.

Mels, T. 1999. *Wild Landscapes: The Cultural Nature of Swedish National Parks.* Meddelanden från Lunds Universitets Geografiska Institutionen, Avhandlingar, 137. Lund: Lund University Press.

Öckinger, E., Lindborg, R., Sjödin, E. and Bommarco, R. 2012. Landscape matrix modifies richness of plants and insects in grassland fragments. *Ecography*, 35(3), 259–267.

O'Rourke, E. 2005. Socio-natural interaction and landscape dynamics in the Burren, Ireland. *Landscape and Urban Planning*, 70(1–2), 69–83.

Peterson, A. 2006. *Farms between Past and Future*. Doctoral thesis, Acta Universitatis Agriculturae Sueciae, 2006(17), 92.

Piqueray, J., Bottin, G., Delescaille, L.M., Bisteau, E., Colinet, G. and Mahy, G. 2011. Rapid restoration of a species-rich ecosystem assessed from soil and vegetation indicators: The case of calcareous grasslands restored from forest stands. *Ecological Indicators*, 11(2), 724–733.

Pykälä, J. 2000. Mitigating human effect on European biodiversity through traditional animal husbandary. *Conservation Biology*, 14(3), 705–712.

Pykälä, J. 2003. Effects of restoration with cattle grazing on plant species composition and richness of semi-natural grasslands. *Biodiversity and Conservation*, 12(11), 2211–2226.

Queiroz, C., Beilin, R., Folke, C. and Lindborg, R. 2014. Farmland abandonment: Threat or opportunity for biodiversity conservation? *Frontiers in Ecology and the Environment*, 12(5), 288–296.

Rotherham, I.D. 2013. Cultural severance and the end of tradition, in *Cultural Severance and the Environment: The Ending of Traditional and Customary Practice on Commons and Landscapes Managed in Common*, edited by I.D. Rotherham. Dordrecht, The Netherlands: Springer, 11–30.

Saltzman, K., Head, L. and Stenseke, M. 2011. Do cows belong in nature? The cultural basis of agriculture in Sweden and Australia. *Journal of Rural Studies*, 27(1), 54–62.

Stenseke, M. 2006. Biodiversity and the local context: Linking seminatural grassland and their future use to social aspects. *Environmental Science & Policy*, 9(4), 350–359.

Stenseke, M. 2016. Integrated landscape management and the complicating issue of temporality. *Landscape Research* 41(2), 199–211.

Stenseke, M. and Berg, Å. 2007. From natural pasture land to "preservation site", in *Food, Raw Materials and Energy: A Knowledge Journey in the Spirit of Linnaeus*, edited by M. Starendal. Stockholm: Formas, 51–56.

Suding, K.N., Gross, K.L. and Houseman, G.R. 2004. Alternative states and positive feedbacks in restoration ecology. *Trends in Ecology & Evolution*, 19(1), 46–53.

Sui, D.Z. 2012. Looking through Hägerstrand's dual vistas: Towards a unified framework for time geography. *Journal of Transport Geography*, 23(3), 5–16.

Swedish Board of Agriculture. 2000. *Om stöd för miljövänligt jordbruk*, SJVFS 2000:132. Jönköping.

Swedish EPA. 2012. *Sweden's Environmental Objectives – An Introduction.* Brochure.

Swedish EPA. 2013. *Specifications for a Varied Agricultural Landscape.* Accessed 21 May 2015. Available at http://www.naturvardsverket.se/en/Environmental-objectives-and-cooperation/Swedens-environmental-objectives/The-national-environmental-objectives/A-Varied-Agricultural-Landscape/Specifications-for-A-Varied-Agricultural-Landscape/.

Swedish Government. 2001. *Svenska miljömål – delmål och åtgärdsstrategier.* Rskr. 2001/02:36.

Swedish National Heritage Board. 2014. *Biologiskt kulturarv – växande historia.* Accessed 16 June. Available at 2015. Available at http://samla.raa.se/xmlui/bitstream/handle/raa/7731/Varia%202014_37.pdf?sequence=1.

Tscharntke, T., Tylianakis, J.M., Rand, T.A., Didham, R.K., Fahrig, L., Batáry, P., *et al.* 2012. Landscape moderation of biodiversity patterns and processes – eight hypotheses. *Biological Reviews*, 87(3), 661–685.

UNESCO, 2015. *Biosphere Reserves.* Accessed 18 August 2015. Available at http://www.unesco.org/new/en/natural-sciences/environment/ecological-sciences/biosphere-reserves/.

Wästfelt, A., Saltzman, K., Berg, E.G. and Dahlberg, A. 2012. Landscape care paradoxes: Swedish landscape care arrangements in a European context. *Geoforum*, 43(6), 1171–1181.

Widgren, M. 2004. Can landscapes be read? In *European Rural Landscapes: Persistence and Change in a Globalising Environment*, edited by H. Palang, H. Sooväli, M. Antrop and G. Setten. Dordrecht, The Netherlands: Kluwer Academic Publishers, 455–465.

Wylie, J. 2007. *Landscape.* London, UK: Routledge.

Index

Milton Keynes UK
Ingram Content Group UK Ltd.
UKHW040103071024
449327UK00019B/787